ISBN: 978-1-312-50537-7

I0475891

GNU Free Documentation License

Published by Lulu Distribution (www.lulu.com)

Printed with pride in the USA.

Eco-Friendly Living

James P. Hewitt, RN

Table of Contents

"Adopting a new healthier lifestyle can involve changing diet to include more fresh fruit and vegetables as well as increasing levels of exercise." - Linford Christie

Introduction

This book has been written with a minimal amount of medical terminology and jargon. My goal has been to create a book that will educate anyone without them needing an encyclopedia and dictionary to make sense of the contents of this book. This book is designed for use as an educational product for the regular person.

As the world becomes smaller, due to globalization it becomes increasingly easy to see how the lives of people, plants, animals, and ecosystems need to be in harmony with one another. What happens on one continent can and does affect all the other continents. There are a multitude of books, videos, and internet resources available regarding going green. This book will attempt to consolidate the information into one easy to understand resource. Much of the information in this book can applied globally.

People are always looking for ways to save money. In today's modern world economy, it is also important to save money, while helping to save the planet. The world is like our bodies. It can only take so much abuse.

The benefits of using safe and more natural products in our lives will help over our lifetime provide a life that is more comfortable, safer, quieter, and healthier. Natural products contain no fabricated chemicals. Some companies abuse the term "natural". Most companies take the commitment to create non-synthetic products very seriously. Using natural products also helps the environment and world. The manufacture of natural products does not release the same toxins into the environment that traditional manufacturing does.

This book is a guide, not an absolute resource. I have gathering quite a bit of information, but this book is designed to whet your appetite regarding the idea of living an eco-friendly life. It has many great ideas and overall I believe good value for the consumer. Remember to read the labels on all the products that you expose to yourself and your family. Please read on, learn, enjoy the book, and spread the knowledge that you absorb.

The truth is that everything we do has an impact on the planet. This impact can be either good or bad. The best part is that as individual's we have the power to control most of our choices. We can make a difference.

"In an underdeveloped country, don't drink the water; in a developed country, don't breathe the air." - Changing Times Magazine

Dedication

This book is dedicated to the men and women who have dedicated their life's work to saving the environment. Nature conservation is a movement that seeks to protect natural resources including animal, fungus, and plant species as well as their habitat for the future. This book is dedicated to you and the good work you do. Humanity appreciates the long hours, the sacrifices, and the dedication you have for our planet.

All of the money raised from the sale of this book will be donated to environmental causes.

Yes, 100% of the money is donated.

Indoor Air Quality

One person alone cannot fix outdoor pollution on their own, but as an individual they can definitely detoxify where they work and live. People spend approximately 90 percent of their time indoors. Many people face risks to their health from pollution. These risks may be greater indoors, due to exposure from indoor air pollution rather than pollution from the outdoors. Research has found that in homes across America, the quality of indoor air can be far worse than outdoor air. Many homes are being built and remodeled tighter, without considering the factors that assure fresh and healthy indoor air. Our homes today use furnishings, combustion appliances, and household products that can compromise the quality of the air.

Aristotle Statue

As early as 384 BC, respiration, which is the process of breathing, was being studied due to its importance to life. Aristotle, an ancient Greek philosopher believed that "we breathed so as to provide a cooling current of air for the fire in our hearts." Take a moment and think of breathing as a vital process, where respiration is vital to life. The air we inhale is composed of roughly 78% nitrogen, 21% oxygen, 1% argon, helium, carbon dioxide, some other gases, and 1% water vapor.

Let us talk about ventilation. The average adult breathes in over 3,000 gallons of air on a daily basis. The normal respiratory rate for an adult is 12 to 20 breaths per minute. Children breathe even more air per pound of body weight. The normal respiratory rate for a child is 20 to 30 breaths per minute. An infant's respiratory rate can be anywhere from 30 to 60 breathes per minute. Children and the elderly are more susceptible to air pollution.

As previously mentioned we often do not think about the air we breathe, even though we breathe often. Most people only think about air quality when there is an odor in the air. An example of this is when you are driving behind a truck and breathe in the diesel exhaust fumes. After a while, most people feel nauseous. During the last several years, scientific evidence has indicated that the air within our homes and buildings can be more seriously polluted than the outdoor air.

Here are some great examples of everyday activities in your home that lead to poor indoor air quality. The majority of people does cooking on the stove, spraying bugs with insecticide, plugging in a room freshener, and shampooing the carpets with detergent.

All of these household activities release chemicals that permeate your home like invisible dust devils waiting to harm your health.

The indoor environment unfortunately is a creation of modern times. Previously, all structures, such as work buildings, churches, and homes were noted for the extent to which they were open to the outside air. Insulation was often inadequate, if there was any at all in the building, depending on its age. Windows were often of poor construction and were designed as a thin single pane. This made them quite drafty, which allowed air to move in and out of the building. This can be commonly called natural ventilation. Unfortunately as the air moves in and out heat can also escape, which most people associate with throwing money out the window.

Today's modern building construction technological advances have permitted us to seal buildings tightly. We now have problems related to our technology advances. Because of these modern advances, there is recirculation of the air within them. Very little outside air may enter buildings through the Heating Ventilation and Air Conditioning (HVAC) systems. Plus, as an added bonus, we fill the buildings with a variety of particle and chemical emitting materials and objects during construction, which often subjects people to breathe recycled air, and not the refreshing clean outside air.

Have you ever walked into a home and felt that the air was stagnant or musty? Does the air feel heavy, moist, or foul? Is your nose stuffed up or runny after a short time in the home? Do you find that certain buildings or dwellings cause you more breathing difficulties? Do they affect your already diagnosed breathing illnesses, such as Asthma, Chronic Obstructive Pulmonary Disease (COPD), or Bronchitis? When you go to open the door to leave a house does the air rush into the house so fast that it makes a whistling sound between the door and the doorframe? Is it tough to open the door to enter the home, because the air rushing into the home is creating a vacuum and the door feels stuck?

During the winter season, your home is like a cocoon. Unfortunately, this airtight home can be harming your family's health. By shutting the windows all winter, you also shut off the ventilation and air circulation. Keeping the windows airtight, means that you keep the house warm, but this will allow the irritating and sometimes toxic particles to build up in the air you breathe.

I was in a high rise in Manhattan and while riding in the elevator heard a high-pitched whistle. The whistle was from the air rushing through the elevator shaft. The shaft was acting like a chimney. Yet unlike a normal chimney, the elevator shaft was allowing the air to be sucked into the building. Think for a minute, have you ever experienced such a phenomenon? These are important questions to ask yourself if you find that certain houses or buildings make you feel ill or affect your health. If you can answer yes to any of these questions, your home may be making you sick.

According to an American Lung Association's State of the Air Report, "Approximately 160 million Americans are breathing unhealthy air each year. Children and seniors are the age groups that are most at risk. In addition, 29 million of these Americans are under the age of 14. Another 15 million are over the age of 65." The Environmental Protection Agency ranks poor indoor air quality among the top five environmental risks to public health. Interestingly, more than half of all Americans are not even aware of this fact.

www.123rf.com

Over half of the United States population lives in communities that have unhealthy levels of either ozone or particle pollution. Unlike outdoor air, indoor air is recycled repeatedly. This recycling causes the air to trap and build up pollutants. It has been estimated that contaminated air results in medical costs of about $1 billion dollars a year.

Removing pollutant sources through the modification or routine maintenance of HVAC systems is important. If your home is damaged by water, the replacement of water-stained ceiling tiles and carpeting will prevent mold and fungal growth. Venting contaminant sources emissions to the outside is primary for keeping the internal air clean. Try to use specific pollutant sources such as paints, adhesives, solvents, and pesticides in well-ventilated areas.

Disease from smoke, pest allergens, mold, radon and other gases is avoidable. The current guidelines of the American Society of Heating, Refrigerating and Air Conditioning Engineers recommend a minimum air exchange rate of 35 percent per hour for houses. Homebuilders have resisted attempts to specify mechanical means to achieve the exchange rate. To do so would cost more and hurt their profit margins. I guess the health of the customer is not important to them.

Fresh air is required in every home to alleviate odors, to provide oxygen for breathing, and to increase thermal comfort. Natural ventilation systems utilize the differences in pressure to move fresh air through structures. These pressure differences can be caused by wind or by the effect created by temperature or humidity differences. The amount of natural ventilation will depend critically on the size and placement of openings, such as windows and vents in the structure. I often imagine a natural ventilation system as a circuit, where equal consideration should be given to intake and output.

Air filters can improve indoor air. Air filters in central heating and air-conditioning ducts in your home or in portable room air cleaners help remove these indoor pollutants. Air filters are used in most central heating and air-conditioning ducts and portable room air cleaners. So, what kinds of air filters are available? There are mechanical, HEPA, ULPA, electronic filters, pleated filters, hybrids or combination, gas phase, and ozone generating filters available to consumers.

Mechanical filters force air through mesh that traps allergens. The group of mechanical filters comprise both the HEPA and pleated filters. HEPA (High Efficiency Particulate Air) and ULPA (Ultra-Low Penetration Air). These filters are usually attached to a furnace. They also can be found as a part of a vacuum cleaner. High-efficiency pleated filters fit in most furnaces and can be either disposable or washable.

Electronic filters use electrical charges to trap pollutants. They charge the air pollutants and these charged particles are then attracted to the oppositely charged filter. Opposites do attract. Hybrid filters contain elements of both mechanical and electronic filters.

Gas phase filters remove odors and gases, such as cooking gas, gases given off by paint

or building materials and perfume. This type
is often used in an automobile body shop.
They cannot remove pollutants such as dust
and pollen.

Ozone generators are touted by manufacturers
as a kind of air cleaner. When inhaled, ozone
can damage the lungs and ozone generators can
be harmful to people with asthma. The ozone
from these generators can irritate the lungs.
Most experts do not recommend the use of
these and neither do I. Remember, getting rid
of the cause of any indoor allergens is the
best way to improve air quality.

Some scientists feel and point out that
exposure to indoor airborne chemicals are
low. Many scientists say that exposure to the
chemicals is cumulative in humans, who have
lived long in the environment and their
health is insufficiently studied. Once
thought to be inert or inactive, many of
these chemicals are now turning up in the
human body. They have even been found in fish
and wildlife.

Breathing is life, so when breathing
improves, all good possibilities in life
hopefully will improve. ALL diseases whether
directly or indirectly are caused or worsened
by poor breathing. How long can you hold your
breath before passing out? Go ahead time
yourself. See if you can hold your breath for
even one minute. Your body cannot live
without breathing, which is indeed
ventilation.

Ben Franklin was a brilliant Scientist, who believed that colds were caused by microscopic creatures and not cold temperatures. The theory during his lifetime was that colder temperatures caused people to become sick.

Benjamin Franklin

Ben Franklin, to prove his theory, slept with the windows open in his bedroom in the clean fresh cold winter air. He actually was less sick than those who did not believe his theory and kept their windows closed and stayed bundled up. We can learn from Ben Franklin.

During the winter season, your home is like a cocoon. Unfortunately, this airtight home can be harming your family's health. By shutting the windows all winter, you also shut off ventilation and air circulation. Keeping the windows airtight, means that you keep the house warm, but this will allow the irritating and sometimes toxic particles to build up in the air you live and breathe. I always crack a window about a ¼ of an inch in my bedroom. Yes, heat escapes, but it also takes stale and musty air with it and lets in cool clean air. Try it and you will most likely find that you sleep better and are sick less often.

Since I wrote my first book, **Sick Building Syndrome**, a growing body of scientific evidence has indicated that the air within homes and other buildings can be more polluted than the outdoor air. This can even occur in even the largest and most industrialized cities, such as New York or Los Angeles. Research shows that people spend the majority of their time indoors. Some people when polled spent 90% of their time indoors. This includes time at home, school, work, and or play. This means that the risks to health may be greater due to exposure to air pollution indoors than outdoors.

"Poor indoor air quality can cause or contribute to the development of chronic respiratory diseases such as asthma and hypersensitivity pneumonitis. In addition, it can cause headaches, dry eyes, nasal congestion, nausea and fatigue. People who already have respiratory diseases are at greater risk." - American Lung Association Indoor Pollution Fact Sheet August 1999

Molds

Ten Fun Facts about Mold!

1. Outdoors, mold exists nearly everywhere.
2. There are over 10,000 species of mold.
3. Mold helps break down, or decompose, organic material.
4. Every home has mold, somewhere.
5. Not all mold is harmfully toxic but it can certainly cause allergy problems.
6. Stachybotrys, a form of toxic mold, is actually uncommon and usually found only in homes that have been flooded or have other severe, prolonged water damage.
7. One common antibiotic is actually a purified mold – penicillium notatum, or penicillin.
8. Mold growth in a home is not limited to flooding & water damage – any exposure to moisture such as a lawn sprinkler, condensation, steam (as from an unvented shower in the bathroom) or periodic leaking that is not attended to will foster the growth of mold.
9. Your homeowner insurance may specifically exclude coverage for damage directly related to mold.
10. Mold spores can not be eliminated from indoor environments; BUT they will only germinate into mold in the presence of moisture.

Ten Facts comes from Homeowner Stuff written by Karen Rice.

A study by the Institute of Medicine in 2004 concluded in several European countries, Canada and the United States indicated that at least 20% of all buildings had one or more signs of dampness. In this study, dampness was defined because of self-reported indicators, such as water leakage or damage, bubbles or discoloration of floor coverings, and visible mould growth indoors on walls, floors or ceilings.

Molds are types of fungi. They grow in the natural environment. Tiny particles of molds are found everywhere even in indoor air. In nature, molds help break down dead materials, and can be found growing on soil, foods, plants and other items. Molds are part of the nature. Molds exist everywhere on our planet. Many different mold species vary enormously in their tolerance to temperature and humidity extremes. Molds can survive harsh conditions such as extremely cold temperatures; highly acidic solvents, anti-bacterial soaps, and can even grow in petroleum products such as jet fuels and heating oils.

Mold is one of the largest biological contaminators. Mold appears as a woolly or furry growth when seen. It is often greenish or whitish in color. It usually appears on food, and other animal, or vegetable substances. Most people have witnessed mold on bread that has gone bad. This happens when they are left too long in a warm, moist place or when they are decaying. Mold is also known as a fungus. Dictionary.com defines mold as, "A growth of minute fungi of various kinds, esp. those of the great groups Hyphomycetes, and Physomycetes, forming on damp or decaying organic matter".

Molds, while not as scary sounding as bacteria and viruses can create harmful health problems for people. The study of molds is called mycology. What are molds? All molds are fungi, but not all fungi are molds. There are roughly about one million species of fungus. Unfortunately, less than one thousand have been studied with regard to human illnesses.

Molds are in simplistic terms a growth of minute fungi of various kinds. Molds produce tiny spores for reproduction. Mold spores drift through the air on a continuous basis. When the mold spores land on a damp spot, they begin growing. Mold only needs a few things to grow and multiply. They only need nutrients, a suitable place to grow, and moisture. Many building materials (such as wood, sheetrock, etc.), paper, carpet, and foods provide food that can support mold growth. When a mold grows, it is known as a colony. When a colony grows large enough, it can be seen by human eyes. They digest whatever they are growing on so they can survive. They also require moisture besides a food source to live. Mold will occur if a moisture problem remains undiscovered or un-addressed.

So, how can you tell if you may have mold? You should use the four of the five senses that you have. You should look, listen, feel, and smell for mold.

Moldy White Bread

I recommend that you do a visual inspection.
Look for visible mold growth. Look around for
any signs of moisture or water damage such as
water leaks, standing water, water stains,
condensation, etc... Examine the areas around
air handling units for standing water.
Routinely inspect the evaporator coils, liner
surfaces, drain pans and drain lines.

Listen for any leaks in your dwelling. Is
there a dripping sound from condensation on
pipes in your basement? Is there a dripping
sound underneath a drain? Either of these are
great ways for mold to grow.

When a person feels ill, they may not
immediately consider the cause to be mold. If
mold allergic people have some of the
symptoms listed above when in your home, you
may have a mold problem. Many symptoms of
mold illness begin to clear up a few hours
after a person leaves the Mold contaminated
environment.

Lastly, sniff your air. Can you notice and
moldy odors? If you have ever smelled mold
before, you will never forget it. It is the

smell of something rotting. It is a musty, tangy odor like there is something moist in the room. If you can smell an earthy or musty odor, you may also have a mold problem.

There are four kinds of health problems that come from our exposure to mold. They are an allergic illness, irritant effects, infection, and toxic effects. For people that are sensitive to molds, may experience some of the following symptoms, such as nasal and sinus irritation, a dry hacking cough, wheezing, skin rash, watery or irritated eyes.

There is no practical way to eliminate all molds and mold spores in the indoor environment. Air purification and filtering help, but the truly only way to control indoor mold growth is to control moisture. Just remember that not all molds are bad. Some molds are cultivated deliberately for our use. Do you like pizza? Do you enjoy chicken wings with blue cheese for dipping? Do you ever eat cheese on a sandwich? Cheese is milk that has gotten moldy on purpose. Blue Cheese is a great example. The blue color that gives the cheese its name is a mold. Specific molds are also useful in the production of antibiotics. Molds have a natural defense against bacteria.

What is humidity? The weather person always talks about humidity as an outdoor topic, so why should I worry since I am indoors? Humidity in the simplest definition is the amount of water vapor in the air. Here is an interesting saying spoken by Warren Hymer in the 1939 movie Mr. Moto on Danger Island, "It's not the heat, it's the humidity". This statement refers to unpleasantly muggy weather. It informs us that humid air can be significantly less comfortable than drier air at the same temperature. What humidity level is comfortable and acceptable? As humans, we control our body temperature by sweating.

The United States Environmental Protection Agency cites the ASHRAE Standard 55-1992 Thermal Environmental Conditions for Human Occupancy, "which recommends keeping relative humidity between 30% and 60%, with below 50% preferred to control dust mites". At higher humidity levels, sweating is less effective for cooling us, so we feel hotter. On the opposite dry air makes us feel cooler at room temperature, which can lead to discomfort, lower productivity and demands for higher settings on the heater thermostat. When the relative humidity is ideal, the temperatures in homes can be lowered without causing discomfort to people in them.

In 1871, the scientist Joseph Lister by chance noticed that the mold, which grows on cheese and fruit, could make germs weaker. It was not until 1928, when Alexander Fleming produced the first antibiotic, Penicillin, which is naturally produced by a mold. The three main mechanisms of mold infection in humans happens through ingestion, inhalation, or contact with the skin. Unfortunately, molds can also cause us health problems. These problems can be mild to severe. Many people when exposed to molds experience allergic reactions, asthma, nasal stuffiness, eye irritation, wheezing, skin irritation, and other respiratory complaints. Some people can have more severe reactions to exposure to molds. Severe reactions may occur among workers exposed to large amounts of molds in occupational settings.

Joseph Lister

The long-term presence of indoor mold may eventually become unhealthy for anyone. The following people may also be affected sooner or more severely. They are babies, children, senior citizens, people with chronic respiratory conditions, and people with a weakened immune system. A few examples are persons with HIV, undergoing chemotherapy, or organ transplant recipients.

An excellent example of this is farm workers who work around moldy hay. Severe reactions may include fever, shortness of breath, shock like symptoms, and anaphylaxis. Some people with chronic lung illnesses, such as chronic obstructive pulmonary disease (COPD), can develop mold infections in their lungs. Molds can trigger asthma attacks in individuals with a history of asthma related problems. People with asthma should try to avoid contact with molds.

Some other illnesses that can be caused by molds are rhinitis, sinusitis, asthma, and skin problems. Mold can produce some toxic illnesses of the respiratory system, liver diseases, infertility, cancer, Lupus, Sudden Infant Death Syndrome (SIDS), Chronic Fatigue Disorder (CFS), memory loss, Alzheimer's type symptoms, and Fibromyalgia.

The number one way to prevent mold growth is to limit moisture. Mold cannot grow without moisture. Trying to reduce indoor humidity will help to decrease mold growth. Increase air circulation by using fans and by moving furniture from wall corners to promote air circulation. Be sure that your house has a source of fresh air and can expel excessive moisture from the home.

Like us, our homes need to breathe. Venting bathrooms, clothes dryers, and other moisture generating sources to the outside will help keep humidity (moisture in the air) to a minimum. During the warmer seasons, using air conditioners and de-humidifiers can help to dry the air. Carpet can absorb moisture and are an excellent place for biological pollutants to grow. If you plan to install carpeting over a concrete floor, I highly recommend that you spend the money on a high quality vapor barrier over the concrete.

Stachybotrys Chartarum, which is more commonly known as Black Mold, is one of the most dangerous molds human beings can encounter. This type of mold can produce chemicals called mycotoxins, which are extremely dangerous. In the 1990's, the Center for Disease Control (CDC) did a study in Cleveland, Ohio and claimed that there was an association between mycotoxins from Stachybotrys spores and pulmonary hemorrhages in infants. Black toxic mold is a very serious problem and you do not want to attempt to get rid of it yourself. This mold should be handled by a professional.

Black Mold

Please remember to routinely wash and clean the filters on air conditioners. Air conditioners can be a source for moisture. If the filter gets wet, you can have mold growth. At that point, mold would be spread all around thanks to the air conditioner fan. The amount of moisture that the air in your building can hold depends on the temperature of the air. As the temperature goes up the air is able to hold more moisture. This is the reason during the summer that you can see the humidity in the air. The same principle holds true for when the temperature goes down. The air holds less moisture. This is why, in cold weather, moisture condenses on cold surfaces. This moisture can encourage mold to grow. Provide adequate ventilation to maintain indoor humidity levels between thirty to sixty percent. Mold grows best in areas where the humidity is maintained at seventy percent or higher.

I recommend cleaning or changing air conditioning filters monthly. Increasing ventilation and using exhaust fans whenever cooking, dishwashing, and cleaning occur will help vent moisture outside. If mold is found in the home, reducing air moisture alone will not be effective. Dead mold may still because allergic reactions in some people, so it is not enough to simply kill the mold, it must also be removed. You need to first clean up the mold and eliminate all sources of moisture.

The best cleaning solution is simple household bleach. Chlorine bleach is the cheapest and best solution for disinfection and cleaning. It should be diluted in a 10 parts water 1-part bleach concentration. This concentration kills everything that can be found in the majority of homes. Always ventilate the area being cleaned and exhaust the air to the outdoors. Never mix chlorine bleach solution with other cleaning solutions or detergents that contain ammonia because toxic fumes could be produced. I also recommend that you should wear an N-95 respirator, which is available at many hardware stores when you clean up mold. Dead mold may still because allergic reactions in some people, so it is not enough to simply kill the mold, it must also be safely removed.

At a hospital in that, I was working in Sarasota, Florida I saw a sign that read, "No fungus among us. Wash your hands". Therefore, you have reduced the air moisture. Great, now what should be done? Absorbent materials that have been wet, such as ceiling tiles or carpeting that is moldy will need to be replaced. Even if mold is not seen, it is a good policy to replace them anyway. Prevention is cheaper and easier than having to fix the problem once it occurs.

Immediately fix any leaky plumbing or sources of interior moisture immediately. Keep all the drip pans used in your air conditioners, refrigerator, and dehumidifier clean and dry. Always utilize exhaust fans or open windows in kitchens and bathrooms to get rid of steam.

Make sure that any clothes dryers are vented to the outside. Maintain low indoor humidity, ideally between 30-50% relative humidity. Look for mold behind cabinets or pictures on cold outside walls, where condensation can occur. Keep furniture away from outside walls. Check around air conditioners and furnaces for stagnant water. Keep these units serviced with regular cleaning of the ducts and air filters.

A dehumidifier removes excess moisture from the air, which as previously stated will limit the growth of mold and other similar contaminants. Dehumidifiers are rated according to their capacity, which is the number of pints of water removed during a 24-hour period. A dehumidifier works by fan-forcing air over cool coils to remove moisture. The condensed moisture drips from the coils into a container. The drier air leaves through the back of the unit. When you first turn the dehumidifier on, set it to the driest setting possible to stabilize the room's humidity as soon as possible. After a couple of cycles set the unit's automatic humidistat to the desired humidity setting you need.

For maximum efficiency and energy effectiveness, keep doors and windows closed to prevent air loss. Maintain an unrestricted airflow. Keep the dehumidifier at least six inches away from walls and furniture. Always keep the coils and bucket clean. Look for frost on the coils. If frost collects on the coil, the air temperature is too cool to operate the machine. Frost keeps the machine from working properly. Turn the unit off and restart when it thaws or when the air temperature gets above 70 degrees F.

What should you do if you have a natural disaster? Is it possible to prevent mold from growing after a flood, hurricane, or other disaster? According to Aurora Home Inspections, here are five tips to prevent mold after water damage. The effectiveness of these tips will be determined, in part, by how long the water was in your home.

The first way may seem obvious, but it does not always happen. After water damage to your home, you want to immediately disconnect electronics in the affected area and move them to dry ground. In addition, any pieces of furniture that can be moved out of the area should be moved immediately to a clean, dry place.

Secondly, remove all carpeting including the padding. There is a possibility that your carpet can be saved with a good cleaning and heavy disinfecting, but that is not a definite.

Thirdly, now that heavy items and carpeting are gone, remove all of the water, down to the last drop. If you have power, a heavy-duty shop vacuum is a great way to soak up floodwater. If you do not have power or are worried about using electrical items, buckets, towels and mops will have to suffice. If the flooding is overwhelming, consider renting a sump pump from a local hardware store.

Fourthly dry out the affected area thoroughly. This is so important. You do not want to leave an ounce of moisture anywhere, if possible. Use fans (commercial fans are great) and a dehumidifier to speed the process.

Fifthly and finally, disinfect, disinfect, disinfect! After the area has completely dried out (including wood beams, insulation, etc), use a heavy disinfectant in all areas. You also want to disinfect the furniture that was removed.

There are no EPA or other federal limits that have been established for mold or mold spores. If you want to test for mold, I personally recommend that the sampling should be conducted by professionals who have specific experience in designing mold sampling protocols, sampling methods, and interpreting results. You do not want to inhale and mold spores.

"Water looks so innocent much of the time, but with a flood it's not over when it's over. Its consequences go on for months. When people see their homes wrecked, and then the mildew and decay set in and they realize that life as they've known it is turned upside down, this can simply be overwhelming." - James Harrison

Lighting

Throughout human history, light has been something humanity has taken for granted. It is there throughout our lives for most of us, and we assume will be there forever. The definition of light according to encylopedia.com is "The form of radiant energy that stimulates the organs of sight, having for normal human vision wavelengths ranging from about 3900 to 7700 angstroms and traveling at a speed of about 186,300 miles per second". This definition is very scientific and does not really tell us much.

The Sun is our supply of daylight. Sunlight is the main energy source for life on Earth. Daily exposure to sunlight is necessary for good health. Sunlight is the best and only natural source of vitamin D. Sunlight is one of the two most important climatic factors for ecosystems, according to experts from the University of Illinois Extension. Exposure to sunlight also helps humans synthesize vitamin D, which helps the muscular, skeletal and nervous systems work properly. Vitamin D is also involved in modulating the activity of human immune cells. Vitamin D deficiency has been associated with increased risks of deadly cancers, cardiovascular disease, multiple sclerosis, rheumatoid arthritis, and type 1 diabetes mellitus.

Another factor that can affect your health in your home is lighting. Homeowners have always desired good lighting for maximum comfort.

What kind of lights are you using in your home? Are they kept clean? Are they covered in dust? Do they all work? Is the light emitted bright, dull, or too bright? Does it cause a glare on your television or computer monitor? I bet you never realized how important those lights are to you. Most likely, you turn them on in the morning and off at night, without even looking at them. So what is "good lighting"? "Good" lighting simply put is lighting that provides enough illumination so that people can see clearly, but are not blinded by excessive light levels.

So how do you know if you have good or bad lighting? What are the signs or symptoms of poor lighting? How do you feel? The most common complaints that people suffer from poor lighting are eyestrain, eye irritation, vision problems, dry burning eyes, and headaches. Poor lighting can also contribute to other bodily problems. People may complain of stiff necks and aches in their shoulders. These complaints can occur when people adopt poor postures when trying to read something under poor lighting conditions.

Computer monitor screens act as a mirror. They reflect objects, shiny walls, and any light source, which can cause glare. Eye discomfort can result from this glare. Unfortunately, the glare forces the user into an awkward position as they try to avoid the glare in their eyes. When people are exposed to glare, they often will lean forward or backward to avoid the glare. An awkward position can lead to eyestrain and often will cause postural fatigue. It is important to have appropriate energy efficient lighting sources in your home to provide a nice soothing environment.

Light pollution is an unwanted consequence of outdoor lighting. This pollution includes sources such as sky glow, light trespass, and glare. Sky glow is a kind of light pollution, visible by the "glowing" effect seen in the skies over many cities and towns as a dome of light. Light trespass is when light enters areas or premises outside the boundary of the premises to be illuminated. Glare is a light within the field of vision that is brighter than the brightness to which the eyes are adapted. In response to the demand to reduce light pollution, research and development efforts have focused on advancements in technology to design luminaries to efficiently direct light where it is needed.

So, how can eye discomfort be reduced? The dimming of bright overhead lights or the use of filters to diffuse overhead lighting can be helpful. You should cover windows with adjustable blinds to prevent outside light from creating a glare. Avoid high gloss on furniture and paint. Instead, use matte finishes on painted surfaces and furniture. Place your television or computer monitor parallel with overhead lights and angle it away from other lights and windows. Use a desk lamp to provide light for any paperwork.

What can you do to reduce eyestrain? Focusing your eyes on objects at the same distance and angle for prolonged periods can contribute to eyestrain. Occasionally, look off in the distance and let your eyes rest. Try not to focus on anything and blink several times. You can even close them for a few seconds to give them relief. Often people ask me if I believe in the use of an anti-glare screen on a computer monitor is helpful. I am not an advocate, because I feel that anything that gets between the person and their monitor can often compromise the image. It is better to reduce glare through proper lighting and monitor placement. Thanks to modern technology, many of the monitors currently in use are already designed with low reflective screens or anti-glare screens.

In the average U.S. home, lighting accounts for about 20% of the electric bill. Americans can save money and help to protect the environment by installing ENERGY STAR lighting. According to the Natural Resource Defense Council, electric power plants are the largest U.S. industrial source of carbon dioxide emissions, which is the main cause of acid rain and global warming. As demand for electricity increases more fossil fuels, like coal, oil, and natural gas are burned by plants to generate power. Energy efficient lighting is a phenomenal way to reduce the amount of carbon emissions added to the atmosphere. As a bonus, energy efficient lights lower your shopping and utility bills, making them a smart long-term investment.

What are Energy Star lights? Can I just get a normal light fixture and use a compact fluorescent bulb for energy efficiency, or is there something special about the specialty lights? The following information shown below came from the Energy Star website.

The main difference between lights is that an energy star light fixture actually distributes the light more efficiently and evenly than standard fixtures. These fixtures can only accept compact fluorescent bulbs (CFLs). This makes a huge difference, especially if they are permanent fixtures in a home. Since a standard incandescent bulb cannot be installed, it guarantees that the light will save money over the long haul.

Governments around the world have passed measures to phase out incandescent light bulbs for general lighting in favor of more energy-efficient lighting alternatives. The Governmental regulations allow for the sale of future versions of incandescent bulbs as long as they are sufficiently energy efficient.

The easiest way to start saving energy is to change out the light bulbs in your current fixtures as they burn out. Energy efficient bulbs last for years. Yes, they cost more initially, but over the long term, they actually cost less. Retrofitting an existing light fixture to accept CFLs is a worthwhile investment. A benefit is that the light bulb produces less heat, which in the summer will reduce the cooling load on the home. It also

means that the fixture can be air sealed, if
it is accessible from the attic. Air sealing
the fixture will save a tremendous amount of
money during the winter months when the home
is being heated. Older recessed fixtures are
inherently leaky. A substantial amount of
heat will exit your home through these tiny
holes.

LED Light Bulb

Another type of light on the market today is
the LED light. LED stands for "light-emitting
diode." LED's have been around since the
1960's. Thanks to technology advances, LED's
are now being used more frequently in homes
for reading lamps, night-lights and in
recessed light fixtures. With their
increasing popularity, LED lights are now
being made available with bases that fit
inside lamp fixtures. LED lights are often
very small and use less energy than
incandescent lights. LED lighting can be more
efficient, durable, versatile and longer
lasting than incandescent and CFLs.

I found an interesting fact about CFLs on the website. "If every household in the U.S. replaced one light bulb with an Energy Star qualified compact fluorescent light bulb (CFL), it would prevent enough pollution to equal removing one million cars from the road".

CFL

When CFLs first came out, they were terrible. They several minutes to reach full brightness. They emitted a harsh blue light, and did not last very long. Thankfully, light bulb manufacturers have made great strides. Today the bulbs are brighter, emit a more natural light, and last for many years.

Here is some eco-friendly data. Choosing a fluorescent bulb over an incandescent bulb keeps approximately 700 pounds of carbon dioxide out of the atmosphere for the duration of the bulbs' life. In 2005, Americans saved enough energy with ENERGY STAR appliances to prevent the release of greenhouse gas equivalent to emissions from 23 million cars, and thereby cut $12 billion from their utility bills.

	LED	CFL	Incandescent
Light bulb projected lifespan	50,000 hours	10,000 hours	1,200 hours
Watts per bulb (equiv. 60 watts)	10	14	60
Cost per bulb	$35.95	$3.95	$1.25
KWh of electricity used over 50,000 hours	500	700	3000
Cost of electricity (@ 0.10per KWh)	$50	$70	$300
Bulbs needed for 50k hours of use	1	5	42
Equivalent 50k hours bulb expense	$35.95	$19.75	$52.50
Total cost for 50k hours	$85.75	$89.75	$352.50

www.eartheasy.com

To learn more about the benefits of Energy Star lighting and compact fluorescent lighting in general, please visit the Energy Star website at www.EnergyStar.gov. As a consumer, you can find the ENERGY STAR label on everything from computer monitors to lights, even newly built homes.

Spending too much time indoors can affect your mood, energy level and productivity. Many studies have shown that people who spend

more time in the sun live happier and

healthier lives. There are many known benefits of sunlight on mood. It is common for people to refer to a person who has a good mood as a someone with a "sunny disposition". Beside the direct effect on mood, sunlight encourages enjoyable outdoor activities that enhance quality of life.

Seasonal affective disorder (SAD) is considered a mood disorder in which people who have normal mental health throughout most of the year experience depressive symptoms in the winter or summer, due to a lack of sunlight. Symptoms of SAD usually begin in October or November and subside in March or April.

The most common treatment modality for SAD is with "light therapy." Light therapy consists of regular, daily exposure to a "light box," which artificially simulates high-intensity sunlight. This is usual done when the person first wakes up. A daylight or sunlight bulb emits more white and blue lights to mimic natural sunlight. These full spectrum light bulbs produce color temperatures upward of 5,000 Kelvin, much like that of natural sunlight. They are available at most hardware stores.

Pool lighting is another area where you can save money. A swimming pool uses a large amount of energy to run the filtration system and lights. I like Pentair's product line. They offer Energy Star rated pumps and lights for use in your pool. "At Pentair, we are

committed to innovative technologies to

maximize energy efficiencies," said Karl R.
Frykman, President, Water Quality Systems,
Pentair. Frykman also states, "In addition to
manufacturing energy-efficient products, we
work to educate our customers about energy
consumption through our Energy Initiative
Program that shares pool energy efficiency
and ENERGY STAR benefits at utility and
energy conferences throughout the U.S.

*"Light can be gentle, dangerous, dreamlike,
bare, living, dead, misty, clear, hot,
dark, violet, spring like, falling,
straight, sensual, limited, poisonous,
calm and soft." - Sven Nykvist*

American's take water for granted. No country in the world uses more water per person than the United States. We do not have to travel to get it. It is in large supply. It is also always available. Try to imagine your life without running water. Imagine having no water to flush toilets, take a daily shower, wash dishes, and wipe the kitchen clean, or water lawns. Welcome to the third world.

Just so, you understand how important water is, the average person needs to drink about four quarts of water a day to replace the body fluids lost through perspiration, sweat, urinating, and breathing. Water helps regulate and maintain your body temperature, transport nutrients, removes waste products, and moistens your mouth, eyes, nose, hair, skin, joints, and digestive tract. When your water intake is limited, it can result in dehydration, an elevated body temperature, fatigue, decreased performance, and you can be at an increased risk for heat related illnesses.

A typical American uses 80 to 100 gallons of water during the course of a day, according to the U.S. Geological Survey. This is for bathing, toilets, clothes washing, consumption, cooking, and anything else we do in our day. The entire country consumes about 323 billion gallons per day of surface water and another 84.5 billion gallons of ground water.

It seems that America is on a water craze.

Everywhere I go I see people drinking bottles of water. The big question regarding water is, "Which is better tap or the bottled water?" Sadly, when it comes to drinking water your best bet may be neither.

The United States tap water is not as bad as tap water found elsewhere in the world, but these days it comes with trace amounts of prescription medications, caffeine and other unwelcome "additives" that may have dangerous effects on all of us, especially our growing children.

Bottle of Water

One of my main concerns with city water is that chemicals such as chlorine are used for disinfection of microbes. Chlorine is used in city water systems to get rid of very small organisms in your water. The main problem with this is that chlorine is poisonous. I understand that it is better than having harmful organisms in the water, but chlorine is not safe to have it in your water. According to the National Cancer Institute,

people who use water with chlorine in it have

a 90% increase in the risk of many forms of cancers. Other in-organics, including heavy metals, nitrates and nitrites (fertilizer run-off), drug residue and organic chemicals can also be present.

Bottled water is often perceived as a better alternative to water from a faucet. Sadly, bottled water can be just as bad as tap water, if not worse. Bottled water does not have to meet the same rigid standards as tap water, especially if it is sold within the state where it is bottled. As a result, it may contain harmful bacteria, other pathogens, and chemicals. Bottling, transporting, disposing of and recycling water bottles requires enormous amounts of energy, which makes bottled water far worse than tap in terms of its impact.

Some Key Differences Between EPA Tap Water and FDA Bottled Water Rules						
Water Type	Dis-infection Required.	Confirmed E. Coli & Fecal Coliform Banned.	Testing Frequency for Bacteria	Must Filter to Remove Pathogens, or Have Strictly Protected Source?	Must Test for Crypto-sporidium, Giardia, and Viruses?	Testing Frequency for Most Synthetic Organic Chemicals
Bottled Water	No	No	1/week	No	No	1/year
Seltzer Water	No	No	None	No	No	None
Big City Tap Water	Yes	Yes	Hundreds/ month	Yes	Yes	1/quarter
Table 1 of NRDC's bottled water report.						

The National Resources Defense Council conducted a four-year review of the bottled water industry and the safety standards that govern it. This study including a comparison of national bottled water rules with national tap water rules, and independent testing of over 1,000 bottles of water. Their conclusion was that there is no assurance that just because water comes out of a bottle it is any cleaner or safer than water from the tap. In fact, roughly 25 percent or more of bottled water is really just tap water in a bottle.

There is really no way to know what you are drinking when it comes to your water these days unless you test every drop you drink, which is simply not practical. My recommendation to have safe clean drinking water is to do it yourself. What I mean is that you should buy a water filter and clean the water you need at the point of use. In home water filtration is the easiest, most effective, most economical, and most environmentally sound way of providing safe, healthy water for you and your family. Drinking clean water is the best thing we can do for our health and in home filtration is the best way to get it these days.

The major benefit to filtering your own water is that you get water that is cleaner than what you can get from your tap or in a bottle at a fraction of the cost of bottled water. Filtering water at home costs only a few pennies a gallon. It costs between one to four dollars a gallon for bottled water. A heck of a difference in price. You do the

cleaning after the water has gone through the questionable pipes that may add dangerous pollutants. You also help the environment, because there is no need to transport your water thousands of miles away. This saves on the burning of expensive fossil fuels and polluting the environment.

With home filtering, there is no need for plastic bottles, which are costly to produce and ship. Plastic bottles also take a vast amount of energy to recycle if they do not end up in landfills. Plastic bottles also can leach dangerous chemicals in the water they hold. Using glass containers at home and on the go are much healthier and greener alternatives to plastic. There are two main types of water filter systems for your home. They are called point of use and whole house filtration.

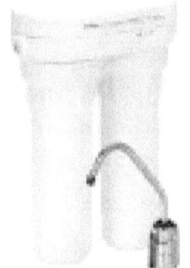

Under Sink Filter

Point of use water filters are installed where the water is to be used and only filter the water at that one point. Point of use water filters include, faucet filters, countertop water filters, refrigerator filters, shower filters, under sink filters, and inline filters.

47

The second option for water filtration and the one I use since my water comes from a well is the whole house water filter. These are usually installed at the point of entry for your water, and use a combination of filtering systems to remove all possible contaminants from your water. Some considerations for choosing a whole house water filter is the needed capacity, output capacity, use of the water, and the cost.

Once you have determined which filter you want, the next question deals with the filtration method. What do you look for in a filter? What are you trying to filter out of your water?

Whole House Filter

Carbon based filters use activated carbon to adsorb lead, chlorine byproducts, certain parasites, radon, solvents, pesticides, herbicides, and some organic chemicals as well as odors and bad tastes. They do **NOT** remove heavy metals, arsenic, nitrites, bacteria, or microbes.

Distiller filters systems boil the water into steam, and then condense it back into water in a separate chamber, leaving behind

particles and total dissolved solids. Since the water is heated, distiller systems do kill microbes. They eliminate many other pollutants including trivalent arsenic, fluoride, lead and mercury. They do not remove volatile organic chemicals and chlorine, which are usually removed by an accompanying carbon filter. Upfront costs for distillers are steep. The prices range between $200 and $1500. They also use considerable amounts of energy. Distilled water contains less dissolved oxygen, and you may find that the water tastes flat. I find this system not eco-friendly as it uses quite a bit of energy to filter your water.

I like ceramic filters. They are often combined with carbon filters. Ceramic filters will remove bacteria, parasites, asbestos and sediments. As the water passes through the pores of the ceramic, particles as small as .2 microns become trapped. When the water flow is reduced, all the filter requires is a light scrub under running water. Ceramic filters are available in both countertop and under the counter models, and are often combined with another filtration method, such as a carbon filter. They use no energy at all and are reusable. They are very eco-friendly.

The last filtration system I will explain is ultraviolet (UV). UV uses an ultraviolet light bulb to kill giardia, E. coli, and Cryptosporidium, and are currently the only systems certified by the National Science Foundation to do so. UV systems are more energy efficient than distillers and are

available in under the counter models and whole house units. UV purifiers are not certified to remove pathogens uncommon to

North America such as toxoplasma or entamoeba and should be used in conjunction with another filtration method. I like to have a carbon filter right after the UV system to filter out any killed off stuff and sediment. I have had this system in my home when I was growing up and it had worked flawlessly for years, with no problems.

Here are some water facts. People getting their water from private wells are responsible for assuring that their water is safe for consumption. The Environmental Protection Agency only requires testing of municipal water sources for about 80 of more than 75,000 known contaminants. The standards for safe drinking water are safe for a 175-pound healthy adult. Those standards may not be safe for children, pregnant and nursing mothers and adults with compromised immune systems.

Municipal water supplies do not need to test for or eliminate many known contaminants that "do not present a health threat". These "secondary contaminants" can make your water taste and smell bad, corrode your pipes, stain your sinks and tubs, may even stain your teeth and alter the color of your skin.

Americans have taken water for granted as a resource. It has been inexpensive and plentiful, and few people have had few incentives to conserve it. Times are

changing. The situation is changing rapidly. Population growth, economic development and prolonged regional droughts are forcing

municipal water systems to seek ways to
reduce the demand for water.

Low Flow Showerhead

Installing low-flow toilets, showerheads, and
aerators for faucets is a simple strategy to
cut water use. Inexpensive and simple to
install, low-flow showerheads and faucet
aerators can reduce your home water
consumption as much as 50%, and reduce your
energy cost of heating the water also by as
much as 50%. This conservation of water and
energy is not only good for the environment,
but the savings in your utility bills will
pay for the cost of the aerators within a few
months.

How much water does your toilet use. Older
style toilets use anywhere from five to seven
gallons per flush. The typical toilet uses
two to four gallons per flush. Thankfully,
there are now low water usage toilets to
avoid all the wasted water.

Early on, low flow toilets got a bad rap.
Their flushing ability left a lot to be
desired. Sometimes it took multiple flushes
to drain and they clogged easily. Today there

are many low flow toilets on the market that
work well. For a toilet to be considered low
flow, it has to meet the current standards

that allow for a maximum water usage of 1.6 gallons per flush.

Showers account for 22% of individual water use in North America. Low flow showerheads help reduce water waste, but do not lower showerhead pressure. A low flow showerhead is a water saving showerhead typically rated at 2.5 gallons per minute or less. There are two types of low flow showerheads. The most common is the aerating showerhead that mixes air with the water. The other is a non-aerating showerhead that does not introduce air into the water stream.

Do you let the sink run while washing dishes? Do you also let the water run while brushing your teeth or shaving? Well, if you do you are wasting a lot of water and energy. Yes, energy. It takes energy to heat that water and pump it to your faucet. Low-flow faucet aerators can cut the water usage of faucets by as much as 40% from 4 gallons per minute to 2.5 gallons. Conventional faucet aerators do not compensate for changes in inlet pressure, so the greater the water pressure, the more water you use. New technology compensates for pressure and provides the same flow regardless of pressure. Aerators are also available that allow water to be turned off at the aerator itself.

To add insult to injury, the containers that many of our canned soups, beans and soft

drinks have been found to contain a controversial chemical called biphenyl A (BPA). This chemical can leak out of the can linings into your food. The plastics industry says BPA is harmless, but a growing number of scientists are concluding, through animal tests, that exposure to BPA raises the risk of certain cancers. Does the plastics industry have your health in their best interests? Most likely, not, so be careful and heed the warning. Plastic water and baby bottles, food and beverage can linings and dental sealants are the most commonly encountered uses of this chemical. BPA has been found to leach from bottles. It moves from can liners into foods, soda, and even from epoxy resin lined vats into wine.

A 2004 bio-monitoring study conducted by the Centers for Disease Control and Prevention (CDC) found that ninety five percent of Americans were found to have the chemical in their urine. Eliminating exposure to BPA is not possible, but there are steps you can take to reduce your family's exposure to this chemical. BPA is found in polycarbonate plastic food containers often marked on the bottom with the letters "PC" recycling label #7. Polycarbonate plastics are rigid, transparent, and used for sippy cups, baby bottles, food storage, and water bottles. Plastics with the recycling labels #1, #2 and #4 on the bottom are safer choices and do not contain BPA.

The US Food and Drug Administration (FDA) strongly reaffirms the safety of food-contact

products containing BPA. The European Food Safety Authority (EFSA) also affirms the safety of BPA in common consumer products.

Canadian Health has also found BPA levels to be safe. Whom to trust? I have a favorite saying, "If in doubt, chuck it out!" My family's health is more important to me. I would not risk their health until BPA is confirmed very safe for ingestion. To avoid BPA you should use a metal water bottle. Be careful as many metal water bottles are lined with a plastic coating that contains BPA. Look for stainless steel bottles that do not have a plastic liner.

I mentioned how Americans have an abundant supply of water, but there are areas of our country were water is scarce. Parts of the Southwest and California are in need of quality water. One method for making potable water that is gaining acceptance is desalination. A major advantage of desalination of ocean water is that water is always available even in the most severe droughts.

Desalination or desalinization refers to the process that removes some amount of salt and other minerals from saline water. Ocean water desalination creates a safe and reliable water supply that is local and not dependent on varying weather conditions or water rights. Ocean-water desalination is the process of removing salt, other minerals and impurities from ocean water so that it can be used to supplement the existing water supply. Desalination is a time-tested process that

originated in the Middle East, dating back to Julius Caesar around 49 BCE.

The first desalination plant built in the U.S. occurred in the 1960's at Guantanamo Bay, Cuba. When water supplies to the naval base were cut off in retaliation for the Cuban Missile Crisis, the base became self-sufficient, desalinating 3.4 million gallons of water every day. As the technology continues to improve, it will become even easier and more affordable to produce desalinated ocean water for everyday human uses.

Some of today's biggest cruise ships, like the Grand Princess of the Princess Cruise Line uses more than 260,000 gallons of fresh water every day. In the past, a cruise ship had to carry potable water, but now it can take from our seas.

In life there seems to be a negative for every positive. Sadly, this is also true with the desalinization. The process results in a fresh water flow, which is the positive, but it leaves a brine by-product, which is the negative. Disposal of the brine flow represents one of desalinization's biggest problems. The brine flow has an extremely high concentration of salt and minerals. The outflow can also alter the temperature and other aspects of the chemical composition of the surrounding ocean water. Such changes are detrimental to marine life.

Here are some interesting facts to ponder.

According to the International Desalination Association, in June 2011, 15,988 desalination plants operated worldwide, producing 66.5 million cubic meters per day, providing water for 300 million people. The largest percent of desalinated water used in any country is in Israel, which produces 40% of its domestic water use from seawater desalination.

The best way in my mind is always done with reducing a person's water usage to lessen humanities impact on our planet.

"Water is the driving force of all nature." - Leonardo da Vinci

Energy Sources

Everyday electricity brings light, warmth,

comfort, leisure and entertainment into our lives. "Solar" is the Latin word for "sun" – and it is a powerful source of energy. The sunlight that shines on the Earth in just one hour could meet world energy demand for an entire year. We can use solar power in two different ways. It can be used as a heat source and as an energy source. People have used the sun as a heat source for thousands of years. Families in ancient Greece used to construct their homes so they could get the most sunlight during the cold winter months to keep them warm.

The Sun is constantly emitting solar energy. It takes roughly 8 minutes for the Suns energy to reach Earth. The solar energy reaching the periphery of the earth's atmosphere is considered constant for all practical purposes, and it is known as the solar constant. This solar constant allows life to grow on Earth. Humanity can use this **FREE** energy to live.

In 1954, scientists at Bell Telephone discovered that silicon, which is an element found in sand created an electric charge when it was exposed to lots of sunlight. A few years' later silicon chips were used to help power space satellites. Today, more than 10,000 American families get all of their electricity from solar power.

The silicon from just one ton of sand, used

in photovoltaic cells, could produce as much electricity as burning 500,000 tons of coal.

Whether on a solar-powered calculator or an international space station, solar panels generate electricity using the same principles of electronics as a chemical battery or standard electrical outlet. With solar panels, it is all about the free flow of electrons through a circuit.

Solar energy is energy from the Sun in the form of heat and light. This energy drives the climate and weather and supports all life on Earth. Heat and light from the Sun, along with secondary solar resources such as wind and wave power, hydroelectricity and biomass, account for over 99.9% of the available flow of renewable energy on Earth. Solar design can provide practical lighting, comfortable temperatures, and improved air quality. This can be accomplishes through the designing of a building's orientation, proportion, window placement, and using appropriate material components for the local climate and environment.

Electricity can be generated from the Sun in several ways. Photovoltaic's (PV) has been mainly developed for small and medium-sized applications, from the calculator powered by a single solar cell to the PV power plant. For large scale generation, concentrating solar thermal power plants have been more common. Thanks to technologies that are more modern new multi-megawatt PV, plants have been built recently. Solar cells convert sunlight to electricity without any moving parts, noise, pollution, radiation or maintenance.

Another great energy source is the wind. Most of us complain about those windy days especially in the winter months, but that wind can provide us with energy. Wind energy is plentiful, renewable, widely distributed, clean, and it helps reduce greenhouse gas emissions when it displaces fossil fuel derived electricity. Wind power is the conversion of wind energy into a useful form, such as electricity using wind turbines. In the past windmills used the wind's energy directly to crush grain and/or to pump water. Unfortunately, humanity only uses the wind currently to produce just over 1% of worldwide electricity use.

The earliest historical reference for a windmill shows that the power of the wind was used to power an organ in the 1st century AD. Windmills were used extensively in Northwestern Europe to grind flour beginning in the 1180s, and many Dutch windmills still exist.

In the United States, the development of the "water pumping windmill" was the major factor in allowing the farming and ranching of vast areas of North America. These areas without the windmills were otherwise devoid of readily accessible water. They contributed to the expansion of rail transport systems throughout the world, by pumping water from wells to supply the needs of the steam locomotives of those early times.

Wind Energy Works

Renewable energy sources like hydroelectric, biomass, geothermal, wind, and solar account for approximately 11% of our electricity supply. Wind power emits no pollution at all. Small wind turbines pose essentially no risk to birds. Small wind turbines do, make some noise and some people believe they cause "visual pollution." Wind power is the fastest growing source of electricity, according to the Electric Power Research Institute (EPRI). EPRI also projects that wind power may be our lowest cost source of electricity within ten years.

A windmill on a farm can make only a small amount of electricity - enough to power a few farm machines. To make enough electricity to

serve many people, power companies build "wind farms" with dozens of huge wind turbines. Wind farms are built in flat, open areas where the wind blows at least 14 miles per hour. One of the largest wind farms in the U.S. is in Altamont Pass, California. It has more than 900 wind turbines.

Wind Turbine

One summer while vacationing on Brigantine Island in New Jersey, I noticed a Wind Farm and Solar Energy Plant near Atlantic City. There are five windmills located on a wastewater site. Here is a blurb from the New Jersey Wind website. "The 7.5-megawatt (MW) Jersey-Atlantic Wind Farm is the first wind farm to be built in New Jersey, and the first coastal wind farm in the United States. The wind farm is located in Atlantic County, NJ and is visible to more than 30 million Atlantic City visitors each year from downtown Atlantic City and the Atlantic City Expressway. The project produces approximately 19 million kilowatt-hours of emission-free electricity per year, which is

enough emission-free energy to power over 2,000 homes. The electricity is used by both

the Atlantic County Utilities Authority
(ACUA) Wastewater Treatment Plant and
delivered to the regional electric grid."

Here are some interesting statistics from
ACUA:
* 500 kW solar electric project makes
  600,000 kW annually.
* Solar panels are on the ground, rooftops
  and parking lot.
* Solar energy project saved more than
  $2,397,752.
* Solar and wind farm prevent over 8,765
  tons of CO_2 in air.

Hydropower is one of the oldest sources of
energy. It was used thousands of years ago to
turn a paddle wheel for grinding grain. Our
nation's first industrial use of hydropower
to generate electricity occurred in 1880,
when 16 brush-arc lamps were powered using a
water turbine at the Wolverine Chair Factory
in Grand Rapids, Michigan.

The first U.S. hydroelectric power plant
opened on the Fox River near Appleton,
Wisconsin, on September 30, 1882. The source
of hydropower is water, so hydroelectric
power plants must be located close to a water
source. It was not until the technology to
transmit electricity over long distances was
developed that hydropower became widely used.

Hydropower accounts for roughly 7 percent of

all U.S. electricity generation and 73
percent of generation from renewable energy
sources. Some people regard hydropower as the

ideal fuel for electricity generation. This is related to the fact that generating electricity via hydropower is almost free. This type of energy production has no waste products, and hydropower does not pollute the water or the air.

Unfortunately, it is often criticized, because it does change the environment by affecting the natural habitats. Many places where Salmon had spawned are now blocked by a dam, but in those places a water ladder has been constructed that allows the Salmon to get to their spawning habitat.

The top states for hydropower production are Montana with five plants, New York with four plants, Oregon with three plants, California has two plants, and Washington State has one. As you can see, much improvement can be done to increase these renewable energy sources. Unfortunately, the creation of new energy plants is expensive. For every positive thought regarding renewable energy, there is also a negative one. Every major renewable

energy source has drawn criticism from leading environmental groups. For example, hydropower for river habitat destruction, wind for avian mortality, solar for desert overdevelopment, biomass for air emissions, and geothermal for depletion and toxic discharges. I have no idea which energy source is best, but the fossil fuel plants are emitting pollution. Their energy source is finite.

The world needs to change, before the lights turn, off permanently.

Hydropower is clean. It prevents the burning of 22 billion gallons of oil or 120 million tons of coal each year. Hydropower does not produce greenhouse gasses or other air pollution. Hydropower leaves behind no waste. The reservoirs formed by hydropower projects have expanded water based recreation resources. In the U.S., hydropower is produced for an average of 0.85 cents per kilowatt-hour. This is 50% the cost of nuclear, 40% the cost of fossil fuel, and 25% the cost of using natural gas.

Heat is naturally present everywhere in the earth. The international geothermal power market is booming, growing at a sustained rate of 5%. Almost 700 geothermal projects are under development in 76 countries. Many countries anticipating the threats caused by climate change realize the values of geothermal power and as source of renewable energy. What is geothermal energy? The word geothermal comes from the Greek words geo

(earth) and therme (heat). Geothermal energy is heat from within the earth. Society can use the steam and hot water produced inside the earth to heat buildings or generate electricity. Geothermal energy is a renewable energy source because the water is replenished by rainfall and the heat is continuously produced inside the earth.

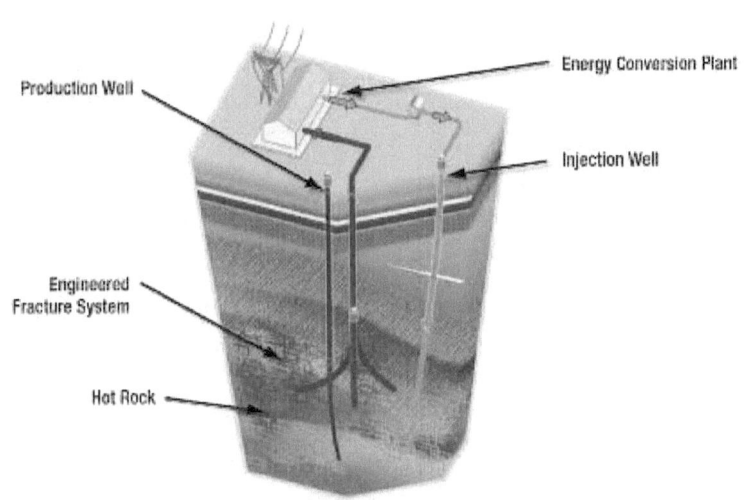

U.S. Department of Energy Geothermal Technologies

The United States generates more geothermal electricity than any other country, but the amount of electricity it produces is less than one-half of a percent of electricity produced in United States. Only four states have geothermal power plants. The first, California has 33 geothermal power plants that produce almost 90 percent of the nation's geothermal electricity. Nevada has 14 geothermal power plants, while Hawaii and Utah each have one geothermal plant. A geothermal power plant is like in a regular power plant except that no fuel is burned to

heat the water into steam. The steam or hot water in a geothermal power plant is heated by the earth. This heated water goes into a special turbine, which is connected to a generator to make electricity. The steam then is cooled off in a cooling tower and is recycled for use again.

The environmental impact of geothermal energy depends on how it is being used. Direct use and heating applications have almost no negative impact on the environment. Geothermal power plants do not burn fuel to generate electricity, so their emission levels are very low and they release less than one percent of the carbon dioxide emissions of a fossil fuel plant. Geothermal plants emit 97 percent less acid rain. Geothermal features in national parks, such as the geysers in Yellowstone National Park, are protected by law, to prevent the land from being disturbed. Native American tribes believe that hot springs and geysers are sacred areas.

Money Isn't All You're Saving

If you are thinking of buying a new household appliance like a dishwasher, clothes washer or refrigerator you need to look for the Energy Star label. Energy Star is a joint program of the U.S. Department of Energy and

the U.S. Environmental Protection Agency designed to certify that certain products meet stringent energy efficiency standards.

The little blue Energy Star labels are a key way to save energy. This helps both the environment and your budget. This program certifies most appliances, with the exceptions of stoves, ovens and clothes dryers. These three appliances do not have a great potential for energy savings. Look for the label on everything from lighting fixtures and windows to microwaves, clothes washers and air conditioners.

Here is an example of how important it is to utilize the Energy Star program. According to the U.S. Department of Energy, refrigerators account for seven percent of a home's total energy use. New Energy Star certified fridges use half as much energy as fridges made before 1993, according to Energy Star. Replacing a fridge made before 1990 with a new Energy Star model would save enough energy to light an average home for nearly four months.

The world is in the early stages of an energy revolution. During the last 150 years, oil has been number one, but the supply is limited. For humanity to survive we will have to evolve our energy use. Energy efficient cars, alternative energies, recycling, organic Bio-fuels, and many other technologies are now becoming a reality, instead of science fiction. It is an interesting time in history to be living. Humanity is going through a multitude of

changes. What used to be science fiction is now becoming a reality.

"Whether fuel cell system development in central Oregon, wind power generation along the Columbia Gorge, or geothermal energy in southern Oregon, investing in new energy sources makes America more energy independent while creating good paying, environmentally friendly jobs." - Greg Walden

68
Odors

The human sense of smell is feeble compared

to that of many animals. Even though, we can recognize thousands of different smells. We utilize our sense of smell for a multitude of activities from enjoying the aroma of freshly brewed coffee to deciding whom not to sit next to on a train. Our noses allow us to detect odors even in infinitesimal quantities, especially if offensive or dangerous.

I am positive that at some point you have been at home and thought, "Something stinks". Unfortunately, the products designed to cover up a smell are usually worse than the original culprit is. The key to clean air is to find and fix the source of the smell. If you handle only the smell, it will keep coming back to linger.

There is a huge difference between an air freshener and an odor counteractant. An air freshener only mitigates the unpleasant odors in indoor spaces, while an odor counteractant does more than mask a smell. It actually gets rid of the odor.

There are a number of reliable general use counteractants, such as Lysol Disinfectant Spray and baking soda. Certain pungent smells definitely call for specific measures. Read on to find the best methods to eliminate those annoying household odors in a safe way. For every nasty scent that you sniff, there is a healthy way to snuff it.

We all love that new car smell. It makes us proud that we have a new car. Whenever someone sits in your vehicle they inhale and

give an "Ah, I love that smell". It gives our
ego a boost. Unfortunately, that wonderful
smell can be bad for your health. The new car
smell comes from toxic gases. Yes, it is
those volatile organic compounds' (VOCS)
again. Your new car is emitting them and you
are being hurt while enjoying that great
smell.

The Ecology Center is a not for profit
environmental organization based in Ann
Arbor, Michigan. They released findings from
a 2006 research project. This report is known
as "Toxic at Any Speed: Chemicals in Cars and
the Need for Safe Alternatives". The Ecology
Center declared that much of the material in
most car interiors that produce that new car
smell is made with toxic chemicals known to
pose major public health risks. The report
detailed that vehicle drivers and passengers
are breathing toxic air. They are also in
constant physical contact with dangerous
chemicals leaching from every interior
surface of the new vehicle.

These dangerous chemicals coat interior
surfaces with a toxic "fog". This is commonly
seen as that new car film seen on interior
windshields and windows. These same dangerous
chemicals are, known to cause birth defects,
impaired cognitive ability, liver problems,
and premature births.

Research on toxic chemical reduction and
elimination has been underway in many

industries for some time. Decanal, formaldehyde, naphthalene and carbon disulfide are used in the production processes of foams, adhesives and fabrics.

The research showed significantly higher levels of VOC's in vehicles as compared to those levels in homes and offices that had been measured in previous studies. This has made in-car pollution a major source of indoor air pollution. That new car smell is definitely a health danger. How can you protect yourself? We all ride in vehicles of some kind, most likely on a daily basis. You can minimize health risks by using solar reflectors. Once again, ventilation plays a key factor. Ventilate car interiors with open windows and air conditioning that does not recycle the interior air.

Car manufacturers have seen the data and are making changes. Toyota has developed an "eco-plastic' derived from sugar cane. Ford Motor Company developed a soy based foam insulation. Ford has also created a bio-fabric for the vehicle interiors. There is a long way to go with these ideas. I hope that these innovations can one day be used in the home and office to help improve indoor air quality.

Clothing can hold odors. Shoes are a great source of a funky smell. Any Mother with teenage boys will agree. When storing clothing due to season changes, avoid

mothballs, as they may be toxic. Mothballs are small balls of chemical pesticide and deodorant. Our grandparents swore by them, but they make your clothing smell like a thrift store. The main threat from mothballs occurs when opening the clothing storage containers, or from wearing clothes immediately after opening. The fumes are a carcinogen.

A solution is to open the containers outside and let the clothes hang and air out for a day before wearing. I prefer to avoid them. The simple solution is to wash your clothes to kill unseemly body odor in fabrics, and store them with an odor remover kit like Dryel, which can help prevent stale odors from storage.

Most Moms would agree that their teenagers' sneakers should be listed as a chemical weapon for the military. The reason for foot odor is related to the bacteria found on our feet combined with our bodily sweat, which causes the odor. It is no wonder our feet stink when you consider the combination of humidity from having your feet encased in sneakers or shoes and the approximately 250,000 sweat glands on each of your feet.

It is a great growing environment for bacteria. To avoid odors, try not wearing the same pair of shoes two days in a row. Let them air out. Avoid those sprays, which can leave a residue and are bad for the environment.

I bet you have opened your refrigerator and

had an offensive smell waft out. Refrigerator can harbor odors from leftover foods to spoiled items. You can spend an afternoon scrubbing your fridge from top to bottom without getting rid of that whiff of food gone badly. Spills often trickle into the drip pan underneath the refrigerator or are stuck onto the door insulation and go undetected for months. There is still nothing better for absorbing fridge odors than good old-fashioned baking soda. Washing the door insulation down with a bleach and water mixture will help kill bacteria and disinfect the insulation. Vanilla extract is also an effective remedy as an odor remover. I like to soak a few cotton ball in vanilla extract and let them sit exposed on a small dish until they dry up.

Ever walk into a kitchen and think, "What smells so good?" On the other side, I bet you have also thought, "What stinks?" The cause for either thought is poorly ventilated or cramped kitchens mixed with pungent foods. Eggs can have a sulfur smell. Cabbage can smell terrible. Burnt foods also are offensive to the senses. Good ventilation is the cure to this problem. Often the best and easiest remedy is to open a window or turn on an exhaust fan to suck the odors from the kitchen.

One smell that drives my wife nutty is the litter box. The ripeness of a cats' litter box depends on the litter as well as on the cats' diet, allergies, or infections. A major

factor is whether the box is cleaned daily.

Cat urine smells awful. If your cat has an accident, the fluids can seep into carpet padding and create nasty smells. If an accident occurs, clean it up right away. I like Spray Nature's Miracle Stain & Odor Remover from my local pet store. It is not harmful to animals or children and it smells pleasant. Clean an area wider than the accident spot to make sure you got all the urine. The best cat litter is the type that is clay based that clumps when wet. This type absorbs odors and is easily scooped. Brushing and bathing your pet can prevent the buildup of animal smells. If you went months without a bath, you would smell also.

Cigarette smoke is insidious. Smoking is the top cause of death and disease that can be prevented. According to the Centers for Disease Control (CDC), smoking causes more than 480,000 deaths each year in the United States. Smoking causes more deaths each year than Human immunodeficiency virus (HIV), illegal drug use, alcohol use, motor vehicle injuries, and firearm related incidents combined.

"More than 10 times as many U.S. citizens have died prematurely from cigarette smoking than have died in all the wars fought by the United States during its history." - CDC

The chemicals in cigarettes cause a number of health risks for smokers and everyone around them. The cigarette odor permeates everything. Scientists have now proven that

there are three types of cigarette smoke

exposure. They are referred to as first hand, second hand, and third hand exposure.

First Hand Smoke is what the smoker inhales directly in to the lungs, while second hand smoke is the smoke that one takes in when next to someone smoking a cigarette. Third hand, smoke or Environmental Tobacco Smoke is a relatively newly discovered concept and most people are not even aware of it. This smoke exposure is the smoke that settles on a smoker's clothing, furniture, floor or carpet and is still present long after the smoker has stopped smoking.

Third hand, smoke is almost impossible to remove. You can always smell when there is third hand smoke lingering around. A great example is when a smoker steps in to your car or you step into to a smoker's car. The smell of the cigarette smoke is present. This is third hand smoke. Scientists have reported that toxins present in second hand and third hand, smoke can cause very severe health complications in babies including sudden infant death.

I do not let anyone smoke in my home as I hate the smell and the habit. If you are

visiting a smoker at their home, try to rely

on cross ventilation by sitting near an open window. Ask them to place a fan in a window, facing out. Smoke free is best, but this is a close second. If they truly enjoy your company, they will be accommodating. Ventilation is the best solution for ridding clothes and furniture of the noxious smell.

I do not know anyone who enjoys the smell of a garbage can. The smells from banana peels, paper towels, cheese scraps, onion ends, milk cartons, and dead flowers is inevitable. Lidded garbage cans left outdoors are especially prone to bacteria growth, due to a warm moist environment.

I like SeaYu Petrotech Odor Eliminator for use in my garbage can. The nontoxic spray binds to odor particles and naturally biodegrades the odor-causing bacteria. Their products are all natural, fur friendly, and child safe. This product is able to stand up to any pet cleanup or odor challenge, even skunk. To prevent garbage smells from forming, clean and disinfect both indoor and outdoor garbage cans with an all-purpose cleaner at least monthly.

Dishwashers are dirty! Sounds funny to think that something that cleans can also be dirty. Most dishwasher odors are caused by dirty plates that have been sitting for a few days. A dishwasher can have odors that is caused by food trapped in the filter or on the tub bottom. If you let your dishes sit before you start the dishwasher, I recommend you run a

"rinse and hold" cycle until you are ready

for a complete wash. To clean the whole machine, run an empty cycle with two cups of white vinegar, which is a natural odor absorber and neutralizer. Remember that your dishwasher is not a garbage disposal. You can prevent odors by rinsing heavy food deposits off before you load and wash the dishes.

My wife and I try to be as organic and natural as possible. We try to create our own air fresheners whenever we can. It is not very hard to do.

The first thing you want to do when creating your homemade air freshener is to think about what scents you love. I personally do not like the smell of cinnamon so we avoid that spice. Always pick scents that make you feel well and comforted. You can even think about the seasons and what you want your home to smell like based on the season.

Fresh ingredients are the only way to go when creating your own personal air fresheners. I enjoy the availability of essential oils. This is where I can have some fun. I like to add any of the following; peppermint, vanilla, maple, almond or coconut extract to many of my air fresheners to create a pleasant scent. If you love tree scents, add some pine needles twigs to places in your living space to natural deodorize your home. Personally, in my opinion nothing smells as great as fresh cut cedar.

Here is a natural room spray recipe. Add 15

drops of the essential oil or a combination
of oils of your choice to a clean one quart
spray bottle. Fill the bottle with 3 cups of
spring water and one cup of vodka. Buy the
cheapest Vodka. The goal is to save money.
Shake it all up to combine the ingredients
and you have a homemade room spray that will
save you tons of money and place no chemicals
into your home. My personal favorites are
lavender, eucalyptus, green apple, and
orange.

Another way to infuse a pleasant scent in a
room is to dampen a cotton ball with the
essential oil of your choice and place in a
dish. I like to hide it so it is not easily
seen. I makes a great quick and easy air
freshener.

Bathrooms are one of the smelliest places in
a home. In the bathroom take a cotton swab
and dampen it with your favorite perfume or
cologne and swab the inside of a toilet paper
roll. Every time the toilet paper is used, a
fresh scent will fill the room.

*"To me, the smell of fresh-made coffee is one
of the greatest inventions." - Hugh Jackman*

Recycling

Human kind is like a cancer to the planet
Earth. We take from it and destroy it every
day. We have ravaged our home. Thanks to
recycling, we may be able to slow down this
destruction and maybe reverse it.

Recycling is very important, as waste has a
huge negative impact on the natural
environment. Harmful chemicals and greenhouse
gasses are released from rubbish in landfill
sites. Recycling helps to reduce the
pollution caused by waste. Habitat
destruction and global warming are some of
the affects caused by deforestation.
Recycling reduces the need for raw materials
so that the rainforests can be preserved.
Huge amounts of energy are used when making
products from raw materials. Recycling
requires much less energy and therefore helps
to preserve natural resources.

Two years after calling recycling a $40
million drain on the city, New York City
leaders realized that a redesigned, efficient

recycling system could actually save the city $20 million and they have now signed a 20-year recycling contract. Thousands of U.S. companies have saved millions of dollars through their voluntary recycling programs. They would not recycle if it did not make economic sense.

Trash comes from many sources, including bottles, boxes, cans, yard trimmings, grass clippings, furniture, clothing, newspapers, and much more. Americans also dispose of several million tons of tires, appliances, furniture, paper, clothing, and other durable and non-durable goods each year as well. Packaging waste, including glass, aluminum, plastics, metals, paper, and paperboard, also contributes significantly to our annual waste totals. Yard trimmings, such as grass clippings and tree limbs, are a substantial part of what is thrown away.

All this trash has to go somewhere. Years ago in college, I worked on a tugboat. We spent time in New York Harbor. Every 11 minutes a group of barges went by loaded with garbage. They were headed out to sea to be dumped. What we have done to our planet is criminal. I even read one story where it was suggested that nuclear waste, instead of being stored on Earth should be shipped into outer space. Not only have we ruined our planet, but also now, we might ruin the universe.

Way to go, Mankind!

Garbage Dump

There is some good news. Americans are embracing recycling programs in records numbers. The EPA estimates that over 30% of the waste produced by Americans ends up in recycling programs. Commonly recycled items include plastics, paper, and cardboard. Communities and businesses have also established recycling programs for some of the more toxic products produced by our society, including batteries, printer/toner cartridges, computers, cell phones, and even used oil.

Many of the plastic products we use today are either recyclable or made of recycled materials. All products made from plastic have a code inside of a triangle on the bottom surface of the container. Have you ever wondered what those codes mean? In the late 1980's, the Society of the Plastics Industry developed a numerical coding system to help indicate which plastic material has been used for a given product. There are six different types of plastic resins that are used in packaging household products.

There are many actions we can take to reduce the amount of waste we generate or that we send to the landfill. You can reuse products. If you do not have a recycling program in your community, or if the material or product is not currently recyclable, try to find another use for the product rather than throwing it away. Composting yard trimmings, food scraps, and other organic wastes can dramatically reduce the amount of waste being sent to landfills. As a side benefit, this composted material makes a great soil for flowerbeds and gardens. You can ask your local grocer to carry paper grocery bags instead of plastic. I always request that my groceries be bagged in paper. Paper products can decompose, unlike plastic. Sell or donate products you can no longer use, such as clothing and furniture, instead of throwing it away.

Recycle, recycle, and recycle!

Take advantage of your community's recycling program. Some organizations estimate that the average family can reduce their weekly waste by 50% through recycling paper, cardboard, cans, plastic bottles, and other recyclable materials. Due to recycling my household has saved money by getting a garbage pickup every other week. My garbage bill is 50% less. Use products made from recycled material whenever possible. Support recycling efforts by purchasing products made from recycled materials.

It takes 95% less energy to recycle aluminum than it does to make it from raw materials. Making recycled steel saves 60%, recycled newspaper 40%, recycled plastics 70%, and recycled glass 40%. These savings far outweigh the energy created as the byproducts of incineration and land filling.

Use rechargeable batteries. Yes, initially they are more expensive, but in the long run as they are reused they truly are much cheaper to use. Using rechargeable batteries not only helps to reduce waste, but it also helps keep the toxic metals found in some batteries out of landfills. A simple way to spread out the cost is to buy the rechargeable batteries as the other ones run out.

Donate or recycle your old computer. Computers contain many materials that are considered toxic and should not be disposed of in a landfill. Rather than throwing out that old computer, see if any local charities, schools, or senior citizen centers can use the computer system. Some agencies will even ship your old computer system overseas to less fortunate people. You save the environment and get a tax deduction.

Here are some recycling facts. I bet you did not know that one recycled tin can would save enough energy to power a television for 3 hours. Also one recycled glass bottle would save enough energy to power a computer for 25 minutes. That soda bottle if recycled would save enough energy to power a 60-watt light bulb for 3 hours. Seventy percent less energy is required to make paper from recycled paper compared with making it from raw materials.

So how can I organize my recycling without having my kitchen look messy and smell foul? In the kitchen, use one container for all recyclables. Rinsing all your recyclables before tossing them into the recycling can will help reduce any odors. If you have curbside recycling pickup, you are probably set. If you have to drop off your own recyclables, then you will also need to create a sorting station. I like having four bins set up for my recycling station. They are for paper, metal, glass, and plastic. Set up your sorting station somewhere you have room, like a garage. Stacking bins keep recyclables tidy. Having wheels, helps make the bins mobile, for easier trips to the curb. My bins cost ten dollars each from www.containerstore.com. No, this is not an ad. You can use bins from any store.

Here are some interesting recycling facts from recycling-revolution.com. Every ton of paper that is recycled saves 17 trees. Recycling benefits the air and water by creating a net reduction in ten major categories of air pollutants and eight major

categories of water pollutants. A national
recycling rate of 30% reduces greenhouse gas
emissions as much as removing nearly 25
million cars from the road. Recycling
conserves natural resources, such as timber,
water, and minerals. Recycling prevents
habitat destruction, loss of Bio-diversity,
and soil erosion associated with logging and
mining.

The bottom line is recycling benefits both
the economy and the environment. What
benefits the economy and environment benefits
all of humanity. Small steps add up to bigger
ones.

Photo taken Lake Placid, New York by Seamus Hewitt.

In 2007 the United States recycled and
composted 85 million tons of the 254 million
tons of municipal solid waste created.
Recycling is only one of the ways you can
help reduce waste and in turn help the
environment.

Below are some of the latest facts about recycling from Epa.gov:

- Each person creates about 4.7 pounds of waste every single day
- In the US 33.4% of solid waste is either recycled or composted, 12.6% is burned in combustion facilities and 54% makes its way into landfills
- In 2007 99% of lead acid batteries were recycled, 54% of paper and paperboard were recycled, 64% of yard trimmings are recycled and nearly 35% of metals were recycled
- The amount of recycling in 2007 saved the energy equivalent of 10.7 billion gallons of gasoline and prevented the release of carbon dioxide of approximately 35 million cars
- The number of landfills in the US are decreasing while their size is increasing. In 1998 there were 8,000 landfills but only 1,754 in 2007
- Each ton of mixed paper that is recycled can save the energy equivalent to 185 gallons of gasoline
- Approximately 8,660 curbside recycling programs exist in the United States
- There are about 3,510 community composting programs in the United States
- Disposal of waste to landfills has decreased from 89% in 1980 to 54% in 2007
- Recycling 1 ton of aluminum cans conserves the equivalent of 1,665 gallons of gasoline

"If you're not buying recycled products, you're not really recycling." - Ed Begley, Jr.

<u>HELP RECYCLE!!!!</u>

Non-Toxic Cleaning

According to a report in the NY Times on September 29, 1997, "Cancer rates have continued to increase every year since 1970. Brain cancer in children is up 40% in 20 years. Toxic chemicals are largely to blame." Household cleaners create toxic waste in their manufacture and use, which gets disposed of in the environment in the form of air and water pollution and solid toxic waste. Not only does this pollution come back to haunt our own health, but it also harms every living thing on earth.

In the Unites States, one in three people suffer from allergies, asthma, sinusitis or bronchitis, according to the US National Center for Health Statistics. The treatment for these conditions should include reducing synthetic chemicals in the home environment.

Therefore, you are in the mood to detoxify your home. This does not mean getting rid of germs has to be overkill. You live in a home, not a hospital. The US Poison Control Centers reports, "cleaning products were responsible for nearly 10 percent of all toxic exposures reported to the U.S. poison control centers, accounting for more than 206,000 calls, over half of which concerned children under the age of six." Antibacterial soaps are helping to promote growth of resistant bacteria, according to a 2000 World Health Organization report. When you use an antibacterial product, you are killing off the beneficial bacteria along with the bad.

The first step in being green is to open a window and let all the pent up pollutants out! I do not understand some people who use air fresheners during the spring and summer, when a vase of lilacs can fill a room with a lovely natural scent. Many consumers stubbornly keep using synthetic room fresheners and fragranced cleaning products that are full of VOCs and other toxic chemicals. These fragrances can make our indoor air unhealthy, provoke skin, eye, and respiratory reactions, and harm the natural environment. They are truly killing you!

Make sure to keep all homemade formulas well labeled, and out of the reach of children.

Consumers need to be alert to what is called "green". Just because a product says, its natural does not mean its nontoxic. The word "natural" is undefined and unregulated by the government and can be applied to just about anything under the sun, including plastic, which comes from naturally occurring petroleum. Unfortunately, since no set standards exist, claims such as "nontoxic," "eco-safe," and "environmentally friendly" are meaningless. Unfortunately, it is a "buyer beware" mentality. I recommend that before you buy, look at the labels for specific eco-friendly ingredients and not harmful ones. If you cannot pronounce the ingredient, it most likely is not healthy and safe.

A few safe and simple household ingredients such as plain soap, water, baking soda, vinegar, lemon juice, and borax can satisfy most household cleaning needs, while saving you money at the same time. Here is a grouping of common, environmentally safe products that can be used alone or in combination for a wealth of household applications.

Baking soda is useful as a cleanser, deodorizer, water softener, and for scouring. Unscented soap is biodegradable and it will clean just about anything. Avoid using soaps that contain petroleum distillates. Lemon, which is one of the strongest food acids, is an effective killer of most household bacteria.

Borax is useful to cleanse, disinfects, softens water, and cleans wallpaper, painted walls and floors.

White vinegar cuts grease, removes mildew, odors, some stains and wax build-up.

Cornstarch can be used to clean windows, polish furniture, shampoo carpets and rugs.

TSP (Trisodium phosphate) is a mixture of soda ash and phosphoric acid. TSP is toxic if swallowed, but it can be used on many jobs, such as cleaning drains or removing old paint. It is useful for cleaning jobs that would normally require much more caustic and poisonous chemicals, and it does not create any fumes.

Here are some companies that seem to be well

known for their eco-friendly household products. I do not endorse any company. They are here as a resource for you, the reader. The list would be enough for a book of cleaners. I am just going to mention a few that I have heard about thru media or personal firsthand knowledge.

The first I will mention is, Seventh Generation. They are committed to becoming the world's most trusted brand of authentic, safe, and environmentally responsible products for a healthy home.

Citra-Solv is dedicated to the idea that people should be able to enjoy high quality cleaning and personal care products without sacrificing either performance or the environment. Citra-Solv specializes in the development of natural products that are made from renewable resources, which have a minimum impact on the planet.

Shaklee Products touts itself as the number one natural nutrition company in the United States. Shaklee has been making people healthier for over 50 years. They also made it part of their job to make the planet healthier along the way.

Vermont Soap produces are USDA approved Certified Organic. Their products replace the often irritating, chemical and detergent based personal care products now in general use. They make handmade cold process bar soaps for sensitive skin, liquid soaps for skin and cleaning. They also offer according to their website, the first truly organic shower gels. They also manufacture numerous organic nontoxic cleaners.

Approach Odor Eliminator has been proven in laboratory tests and in top US and European hospitals to be completely safe to use. It is hypoallergenic, non-toxic, non-irritating, non-staining, fragrance free and environmentally friendly. It is safe for your, children, pets, and any water safe surface in your home. It neutralizes odors rather than covering them up.

The Body Shop is another favorite store for cleansers. They are a leader in promoting greater corporate transparency, and have been a force for positive social and environmental change. As a company they have five core Values, which are to Support Community Trade, Defend Human Rights, Against Animal Testing, Activate Self-Esteem, and Protect Our Planet.

I like to get my scented oils for air freshening and cleaning from Wild Hibiscus. They look for the best of what nature has to offer, which keeps their products vibrant and intense in colors, scents, and flavors.

Another company that I have been told offers exceptional green cleaning products is Nellie's All-Natural. Their natural products provide more than an exceptional clean as their products truly help your family go green, by stripping away toxins and chemicals. Nellie's offers both household as well as laundry cleansers. Their products cleans without leaving behind any residue that can be absorbed into your skin.

My last company to recommend is Green Virgin Products. They are in Tampa, Florida and only sells toxic-free, eco-friendly products. They also use high quality ingredients. What is cool is that they offer a 90-day money back guarantee that their "Green" products work as well or better than leading national brands.

"The business of business should not just be about money, it should be about responsibility. It should be about public good, not private greed." - Anita Roddick

Green Plant Solutions

Think of a space in your home that is a bit barren or unwelcoming. Could a few plants liven up these areas? Any dark corner, nook, or staircase can be brought to life with the appropriate houseplant. As a Registered Nurse I am always amazed at how a sterile looking hospital room instantly, becomes a lot more welcoming with the addition of some healthy greenery.

Research has shown that houseplants can have a positive effect on our mood, reduce our stress levels, and improve our attitude. Diane Relf, a professor of horticulture at Virginia Polytechnic Institute, published an article that mentioned how a major manufacturing company incorporated plants into the interior design of its office space. The design included a standard that no employee in the office area was placed more than 45 feet from any vegetation. The Company administrators found that by adding the vegetation to the work environment helped employees with enhanced creativity and increased employee productivity.

Houseplants have benefits that can improve your life. Houseplants bring the fresh beauty of the outdoors inside your home. They produce the oxygen that makes all life possible. The plants add precious moisture to the air and help to filter toxins. In this chapter, I hope to educate you as to how houseplants can be both beneficial to your health and overall well-being. In addition, in this chapter I will show you how to purify the air and environment with some simple houseplants. Here is a little statistic for you. Americans spend almost 90 percent of their lives indoors. We all work, live, or go to school inside a building. This all means that good indoor air quality is vital for good health.

How do plants help? What is photosynthesis? How can photosynthesis help us have cleaner air? Very simply explained, photosynthesis is the way a plant makes food for itself. Chlorophyll, which is the "green" part of the leaves, captures light energy from the sun. This energy powers the building of food from two very simple ingredients. They are carbon dioxide and water. Oxygen is released as a by-product of photosynthesis. Plants "breathe in" carbon dioxide and "breathe out" oxygen, thus cleaning the air we breathe.

Did you know the air inside our homes and businesses is 10 times more polluted than the air outside? How can you cleanse the air to keep your home or work place a safe haven? A simple and inexpensive answer is to fill your home or work space with houseplants. You will get the best air purification by clustering plants in areas where you spend the most time, such as by your bedroom or on your work table. Houseplants are the best natural filters of common pollutants such as ammonia, formaldehyde and benzene. Hundreds of these poisonous chemicals are released by furniture, carpets and building materials. They are trapped by closed ventilation systems, which leads to a host of respiratory and allergic reactions now called, "sick building syndrome." In the 1980s, NASA scientists studied methods to reduce indoor air pollution. Their research found that houseplants, when grown in a closed, controlled environment, were able to extract volatile organic chemicals from the air.

You are probably wondering, which plants are the best. Read on and find out which houseplants are our most effective allies in keeping your indoor air clean and pure. It is recommended that you should have one plant per 10 square yards of floor space. Essentially this means that you need roughly three plants to contribute to good air quality in the average domestic living space of about 25 square yards. Research done for NASA has shown that there are 10 plants that are most effective all around in counteracting off gassed chemicals and contributing to balanced internal humidity. They are the Areca Palm, Reed Palm, Dwarf Date Palm, Boston fern, Janet Craig Dracaena, English Ivy, Australian Sword Fern, Peace Lily, Rubber Plant, and Weeping Fig.

Rubber Plant

Many plants like light, but they do not all have to be placed near windows. Most indoor plants came originally from the dense shade of tropical forests. These plants also have a high rate of photosynthesis. These plants are ideal for indoor use. When positioning plants, try to strike a balance between light and ventilation. The effect of plants on indoor air pollution appears to be reduced if they are set in a draft. Another benefit of houseplants are that they can be used architecturally and artistically to improve your home décor.

A NASA study that tried to determine which plants would be beneficial in a lunar colony showed that four plants lower formaldehyde concentrations and other impurities in the air, while emitting significant amounts of oxygen. They are Spider Plants, Philodendron, Ivy, and the Snake Plant. The study also found that many plants also lower levels of airborne benzene and trichloroethylene. Houseplants in the office can increase your level of work health by absorbing computer emissions.

Spider Plant

The spider plant is one of the most common houseplants. It is easily grown and is especially popular for the ease and speed with which it forms new plants. Spider plants grow quickly to 2 to 2½ feet wide and 2 to 3 feet long when grown in a hanging basket. The long, grassy leaves are available in green or striped yellow or white. Long wiry stems appear on healthy plants with many small white flowers and miniature plantlets. If these new plantlets touch soil, they will root. These plantlets can be either detached to produce new plants, or left on to create a very full basket.

Philodendrons are among the most common and easy to grow houseplants. Many tolerate low light and neglect. If well treated, they will be beautiful and dependable for many years. Vine types can be limited in height by limiting the height of their support. They also can be limited thru training and pruning. The self-heading types eventually can become very large and should be given ample space to grow. This diverse group of plants ranges from vines with 3-inch heart shaped green leaves to vines with leaves 3 feet long. Some types have glossy solid green leaves; while others have velvet textured patterned leaves. Some even have deep red leaves and stems. Self-headers send out leaves from a heavy clump of growth at their base. These often have dramatically large leaves in a variety of shapes.

Ivy is a genus of 15 species of climbing or ground creeping evergreen woody plants. On suitable surfaces such as trees and rock faces, they are able to climb to at least 50 to 75 feet above the basal ground level. Ivy plants have two leaf types, with juvenile leaves on creeping and climbing stems, and unlobed cordate adult leaves on fertile flowering stems. The Ivy shoots also differ, with the juvenile being slender, flexible and scrambling or climbing with small roots to affix the shoot to the rock or tree. The adult shoots are thicker, self-supporting, and without roots. The flowers are produced in late autumn, individually small. The fruit from the Ivy are small black berries ripening in late winter, and are an important food for many birds, though poisonous to humans.

The Snake plant is the ultimate for those without a green thumb. This plant was perfect for my Mother in Law. Even she could not kill it. This houseplant is one of the hardest to kill. The Snake plant contains heavy, sword like leaves that shoot up from the base. The plant will grow in a clump like style. Another common name for this plant is the Mother-In-Law's Tongue. This houseplant is toxic when eaten. It is one of the many poisonous houseplants.

Approximately 2.5 million poisonings are reported to U.S. poison control centers every year, with nearly 1,000 reported deaths. Here is a list of poisonous houseplants. Not only are these houseplants poisonous to animals such as cats and dogs, but also humans. Please keep your young children away from them. The poisonous houseplants are; Aloe

Vera, Amaryllis, Angels Wings, Anthurium,
Bird of Paradise, Chinese Evergreen, Corn
Plant, Croton, Crown of Thorns, Devil's Ivy,
Dieffenbachia (Dumb Cane), Dracaena Palms,
Elephant Ear, English Ivy, Fiddle-leaf fig,
Fishtail Palm, Gold Dust Dracaena, Heart leaf
Philodendron, Janet Craig Dracaena, Peace
Lily, Poinsettia, Pothos, Ribbon Plant,
Rubber Plant, Sago Palm, Schefflera, Snake
Plants (Mother-in Law's Tongue), Spider
Plants, and the Split Leaf Philodendron.

Standard Poison Sign

Approximately 2.5 million poisonings are
reported to U.S. poison control centers every
year, with nearly 1,000 reported deaths. Here
is a list of poisonous houseplants. Not only
are these houseplants poisonous to animals
such as cats and dogs, but also humans.
Please keep your young children away from
them. The poisonous houseplants are; Aloe
Vera, Amaryllis, Angels Wings, Anthurium,
Bird of Paradise, Chinese Evergreen, Corn
Plant, Croton, Crown of Thorns, Devil's Ivy,
Dieffenbachia (Dumb Cane), Dracaena Palms,

Elephant Ear, English Ivy, Fiddle-leaf fig, Fishtail Palm, Gold Dust Dracaena, Heart leaf Philodendron, Janet Craig Dracaena, Peace Lily, Poinsettia, Pothos, Ribbon Plant, Rubber Plant, Sago Palm, Schefflera, Snake Plants (Mother-in Law's Tongue), Spider Plants, and the Split Leaf Philodendron.

What should you do if you suspect that your family member has possibly eaten a poisonous plant? Seek immediate medical help. Call 911 or your local emergency number immediately if the person is drowsy, unconscious, breathing with difficulty, stopped breathing, or if the person is having seizures.

DO NOT make a person throw up unless told to do so by Poison Control or a qualified health care professional.

If the person seems stable and has no symptoms, but you suspect poisoning, call the Poison Control Center at (800) 222-1222. This is a free and confidential service. All local poison control centers in the United States use this national number. You should call if you have any questions about poisoning or poison prevention. It does NOT need to be an emergency. You can call for any reason, 24 hours a day, and 7 days a week. The National Poison Control Center can be called from anywhere in the United States. This national hotline number will let you talk to experts in poisoning.

Provide information about the person's symptoms and, if possible, information about what he or she ingested, how much and when. Identify yourself. Describe the patient by name, age and gender. Have the container or poison in your hand. Explain how poison was taken. If going to the Doctor's office or Emergency room, bring some leaves of the plant for proper identification by a health care provider.

Most puppies and many adult dogs will chew on plant foliage out of curiosity, boredom, or an attempt to induce vomiting. The trouble is many of the chewed-on plants are poisonous. What do you do for your pet? Whom are you going to call? Do not despair help is available.

The National Animal Poison Control Center, which is a division of the American Society for Prevention of Cruelty to Animals, is only a phone call away every hour of every day of every week. Center veterinarians and veterinary toxicologists have the most current information on toxicity levels, antidotes, and treatments for your pet. If you suspect your pet has been poisoned, gather the following information and then call the NAPCC.

Identify yourself by giving your name, address, and telephone number. They will need to know the species, breed, age, sex, and weight of the affected animal. The substance the animal ingested and how much time has elapsed since ingestion. They will also want you to describe any symptoms that the animal may be showing. The NAPCC's telephone number is (888) 4ANIHELP (426-4435). They charge a small fee called a donation for their help.

Remember that houseplants emit the oxygen we need to breathe while absorbing the carbon dioxide we exhale. They absorb toxins in the air. Plants absorb those VOCs mentioned above often linked to "Sick Building Syndrome". Houseplants can help reduce headaches, sore, dry throats, dry itchy skin, and fatigue. Plants keep us calm. They help us to recover from surgery. They can absorb noise. Remember that the air quality in a workplace is especially toxic, because of poor ventilation systems and commercial cleaning products. Avoid "sick building syndrome" by loading your desk or the area nearby with houseplants.

Save Energy and Money

Having an environmentally safe house also entails saving money. The less energy that you use equates to less energy that has to be made or harnessed. As we already are aware electricity is primarily made thru coal fired plants or nuclear plants. Humanity has started to increase its use of natural renewable energy like wind, solar, and water. I call this the harnessed energy. Some of this chapter will overlap previous chapters, but the information is very important.

Reducing energy costs at home is NOT that difficult. It will require you to make some minor changes in the way you do things. Ultimately, the payoff is reduced energy consumption and lower utility bills. Most people will agree that this is worth the effort. I have tried many of these suggestions and have been amazed at how effective they have been regarding my utility bills. Here are some ways to improve your household's energy efficiency and save money.

My first suggestion is that you make the switch from traditional incandescent light bulbs to compact fluorescent bulbs as soon as possible. I discussed the benefits to these bulbs in an earlier chapter. The compact fluorescent bulbs are more expensive than incandescent bulbs; they use only 25 percent of the electricity and last eight to 10 times longer. My recommendation is to go to your hardware store and buy a couple of the compact bulbs. As a light bulb burns out, replace it with the new compact light bulbs. This will save you a bunch of money initially as you will not have to change all the light bulbs in your home at once.

Always turn off lights and unplug electronic devices when no one is using them. Sounds simple, but how often do you walk into a room and the lights have been left on. In stand-by-mode, the TV, DVD player, kitchen appliances, cordless telephone and chargers for portable devices all still use power.

I use a power strip with a single on/off switch to turn off my home computer station. This turns off several devices in one easy step. My printer, monitor, high-speed router, speakers, and satellite receiver are not utilizing any energy. Look for the Energy Star label when you are buying appliances, home electronics, light fixtures and heating/cooling systems. These household products meet strict energy efficiency guidelines set by the government. They are designed to use less energy, save money, and help protect the environment.

Check with your local utility service and ask if they offer a lower rate at "off peak" times. My energy supplier is NYSEG, which offers a special nighttime reduced cost rate. I operate my dishwasher, clothes washer and dryer at those times. I also do my laundry using cold water and wash only full loads. Clorox bleach now even offers a cold-water formula for those whites. Most of us waste energy heating our hot water. We turn on the hot and then mix it with cold water so we can tolerate it. We should set the temperature so we only have to turn on the hot water to take a shower or bath. Most dishwashers heat up the water for you and the germs are washed away. If you want to go by the older guidelines on hot water temperature, set your water heater at or below 120 degrees. You can save 13 percent for each 10 degrees lower. Keep appliances clean and well maintained. Dryer lint, dusty refrigerator coils and soiled reflector pans reduce energy efficiency.

Most of us like natural light, but in the summer, pull down window shades to block the sun's rays before it heats up a room. This will help keep your house cooler and save on the air conditioner use. At night when there is a cool breeze, open the windows to cool, your home is interior instead of leaving the air conditioner running. Moving air feels cooler to the skin, so consider installing a ceiling or window fan. Thanks to the fan's cooling effect, you can raise the thermostat a few degrees, reducing energy consumption and air conditioning costs.

Do not place a lamp or television near the thermostat for a central air system. This will makes the unit run longer as it will not sense the correct temperature in the house. Clean or replace air filters on your central air conditioning system or window unit once a month. Dirty air conditioners have to work much harder to move the air.

The average American household spends about $750 on heating each winter, but you can lower that bill with a few easy steps. Oil prices are not going down.

In the winter, I recommend that you install a programmable thermostat. A programmable thermostat allows you to automatically lower the heat when no one is home and raise it back up prior to your return. A programmable thermostat can reduce heating costs by about 15 percent. A programmable thermostat typically sells for around $40.

Maintain your heating system. I recommend that you have it serviced annually by a trained technician. Keep your furnace clean, lubricated, and properly adjusted. Clean or replace the filters regularly. If your furnace is more than 15 years old and unreliable, a new energy-efficient model could cut heating bills. Set the thermostat at the highest comfortable level in the summer and the lowest in the winter. For example, a summer setting of 78 degrees uses about 40 percent less power than an air conditioner set at 72 degrees.

Avoid wasting water by installing water saving showerheads and faucet aerators. I also mentioned this in a previous chapter, but water saving aerators are inexpensive devices that restrict the flow of water. Always fix leaky faucets and dripping toilets to avoid wasting water and keeping your water bill lower. If you use a hose and sprinkler to water the lawn or garden, set a timer as a reminder to move them around the yard. If you have an automatic watering system, make sure it is set to turn on and off at the correct times.

There are devices that you can attach to your garden hose that will shut off the water after a programmed time. Whenever your faucet or toilet runs, it costs money. Store a pitcher of drinking water in the refrigerator, instead of letting the faucet run until the water is ice cold.

As much as I advocate for natural ventilation, plugging leaks in the exterior of your house will save heating and cooling costs.

Major Sources of Air Leaks

Floors, walls and ceilings
31%

Fans and Vents
4%

Windows
10%

Electrical Outlets
2%

Ducts
15%

Doors
11%

Plumbing Entries
13%

Fireplace
14%

New England Energy Consulting Company

Small openings in the outer shell of a house account for nearly 30 percent of a home's total heat loss. I like to use silicone based caulk to seal cracks less than ¼-inch wide between door and window frames, openings on exterior walls, around pipes, or anywhere I feel a draft.

For larger cracks, I use an expanding foam sealant. Weather-stripping around doors and windows is another improvement that reduces air leaks.

Organic Foods

Once found only in health food stores, organic food is now a regular feature at most supermarkets. As if reading product labels is not hard enough, now you have to read between the lines to truly figure out what is truly "Organic".

To be an Organic crop they must be grown in safe soil, have no modifications, and must remain separate from conventional products. Farmers are not allowed to use synthetic pesticides, bioengineered genes, petroleum based fertilizers, and sewage sludge based fertilizers.

To be an Organic livestock they must have access to the outdoors, often called free range and be given organic feed. The livestock may not be given antibiotics, growth hormones, or any animal byproducts.

The U.S. Department of Agriculture (USDA) has established an organic certification program that requires all organic foods to meet strict government standards. These standards regulate how such foods are grown, handled and processed.

Any product labeled as organic must be USDA certified with the exception for producers who sell less than $5,000 a year in organic foods. They are exempt from this certification. In though they are exempt they are still required to follow the USDA's standards for organic foods.

The Organic Foods Production Act (OFPA),
enacted under Title 21 of the 1990 Farm Bill,
establishes the uniform national standards
for the production and handling of foods
labeled as "organic." The Act authorized a
new USDA National Organic Program to set
national standards for the production,
handling, and processing of organically grown
agricultural products and the mandatory
certification of organic production.

The Organic Foods Production Act, defines
specific organic standards. These standards
are; preserve natural resources and
biodiversity, support animal health and
welfare, provide access to the outdoors so
that animals can exercise their natural
behaviors, only use approved materials, do
not use genetically modified ingredients,
receive annual onsite inspections, and
separate organic food from non-organic food.

In April of 1995, the USDA National Organic
Standards Board established the following
definition, "Organic agriculture is an
ecological production management system that
promotes and enhances biodiversity,
biological cycles and soil biological
activity."

I live in New York State, which is known for
its apples. New York is the second-largest
apple producing state in the country.
Therefore, I will use the apple as an example
to help you better understand.

Therefore, you go to the supermarket and see

a bunch of apples. There are conventionally grown apples. You also will see some that are listed as organic. Both apples are firm, shiny and red. They both will provide vitamins and fiber. Which one will you choose?

According to the Hartman Group, a market research firm, "About 70 percent of Americans buy organic food occasionally, and nearly one quarter buy it every week." Organic food has become very popular. However, navigating the maze of organic food labels, benefits, and claims can be confusing. Food labels have always been confusing for consumers.

The government tries to make it easier, but to look at the label on a food container you almost need to be a scientist. There are percentages, weight measures, nutrition breakdown per serving, nutritional breakdowns according to a 2000-calorie a day diet. This list of confusion goes on. I have made it simple when it comes to being labelled, "Organic".

Products that contain at least 70 percent organic ingredients may say, "Made with organic ingredients" on the label, but may not use the seal. Products must be at least 95 percent organic to use the term, "Organic" on their label. Products must be either completely organic or made of all organic ingredients to use the label, "100 percent organic".

According to FreeDictionary.com, organic food

is defined as, "Food grown or raised without the use of additives, coloring, synthetic chemicals (e.g., fertilizers, pesticides, and hormones), radiation, or genetic manipulation and meeting criteria of the U.S.D.A. Standard National Organic Program."

According to the Organic Trade Association, "Organic farming reduces pollutants in groundwater and creates richer soil that aids plant growth while reducing erosion."

Certified Naturally Grown farmers, such as "Fledging Crow Vegetables" in Keeseville, NY share a commitment to work within the natural biological cycles that are necessary for a truly sustainable farming system. This natural biological growth system works in harmony with microorganisms, the soil's natural flora and fauna, pollinators, other plants and all animals.

The "Certified Naturally Grown" guarantees that farms do no use artificial treatments or pesticides, and that the farm's crop management maintains or improves the natural environment.

I truly believe that organic food provides a variety of benefits. Studies show that organic foods have more beneficial nutrients, such as antioxidants, than their conventionally grown counterparts do. An added benefit of organic food is for people with allergies to foods, chemicals, or preservatives. These folks often find their symptoms lessen or go away when they eat only organic foods.

My family eats organic food as often as possible. We make this choice to help limit our exposure to synthetic insecticides, fungicides, and herbicides. Many of these chemicals are known carcinogens.

I asked people at a few local grocery stores and natural food markets why they choose organic foods or their feelings about organic foods. Here are some of their replies.

"Organic means fewer pesticides."

"Organic food is fresher."
"Organic farms are better for the environment."

"Organic food is GMO-free."

"I eat nothing that's processed or refined, no high-fructose corn syrup, no sugar, no trans-fats."

"When I can afford it, I'm very into organic food."

"I don't always buy organic food. It is more expensive."

*"Organically raised animals are **NOT** given antibiotics, growth hormones, or fed animal byproducts."*

"It makes me feel like I am doing my part for our planet."

"God made the plants and animals. It is humankind that has screwed them up. I eat like God desired in the beginning."

These are just some of the responses I received. They are all valid responses. Whatever your reason, I know of no one who says eating chemically covered, GMO, or foods filled with hormones are better for your

"I would like to see people more aware of where their food comes from. I would like to see small farmers empowered. I feed my daughter almost exclusively organic food." - Anthony Bourdain

Wash Your Fruit and Veggies

Every year, nearly 48 million people fall ill from food contamination, including sickness caused by fruits and vegetables. Animals, dangerous substances in soil and water, poor hygiene of food employees and other circumstances can lead to food contamination.

Most of us grew up being told to eat our fruits and veggies. My Mother always told me to run any fresh fruits and vegetables under tap water to clean them prior to eating. Now as a Nurse, I know that eating a variety of fresh fruits and vegetables is key to maintaining a healthy diet and overall lifestyle.

I also understand the reason why my Mother always had me was my fresh fruit and veggies. Recently, in the past few years there has been an increasing number of food borne outbreaks related to produce and the growing concern about the use of pesticides on produce, which has made food safety an important issue.

Washing fruits and vegetables not only helps remove dirt, bacteria, and stubborn garden pests, but it also helps remove residual pesticides.

So, is there a proper way to safely wash your fresh fruits and vegetables prior to consumption? The answer is yes. Good food hygiene can have a significant impact on our personal and public health.

Do not wash fruits and vegetables with detergent or bleach solutions. Many types of fresh produce are porous and could absorb these chemicals, thus changing their taste and adding harmful substances to your food.

The FDA advises against using commercial produce washes, because the safety of the washes residues has not been evaluated nor has their effectiveness been tested or standardized. The U.S. Food and Drug Administration does recommend washing produce thoroughly. The easiest way is by soaking the produce in a 1 to 3 vinegar and water mixture. This ensures the acidic blend kills all bacteria.

Allow the produce to rest for 30 seconds before rubbing its surface and rinsing it under cold, running water. Washing berries with a vinegar solution offers additional benefits. It prevents the berries from molding within a few days of purchase.

Cleaning melons is a little different. Melons often have rough netted surfaces which provide an excellent environment for microorganisms to hide. These organisms can be transferred to the interior surfaces during cutting. I like use a vegetable brush and wash melons thoroughly under running water before using the vinegar soak.

Your sink makes a great container for cleaning and soaking your produce as long as the sink is cleaned prior to use and after each use.

Safe Meat Handling

Foodborne illness can strike anyone. Certain people are at higher risk, such as pregnant women, young children, older adults, and people with weakened immune systems.

By the time meat reaches your kitchen, Government inspectors have overseen meat and poultry processing to verify compliance with federal regulations. It is essential that you take steps to maintain safety all the way to the table. Practicing safe-handling methods in the home can reduce the risk of food borne illness and keep your family healthy.

When it comes to keeping food safe the first item to check is your refrigerator. Yes, the fridge. Your refrigerator temperature should be at 40 degrees F or below, to keep foods out of the "danger zone." Keeping foods below 40 degrees inhibits bacterial growth.

NEVER defrost frozen meat and poultry products at room temperature. Keeping the products cold during defrosting is the key to preventing bacteria from growing.

Wash all utensils, cutting surfaces and counters with hot, soapy water after contact with meat and poultry and keep fresh meat and meat juices away from other foods, both in the refrigerator and during preparation.

Wash hands thoroughly in hot, soapy water before and after handling meat and other fresh foods.

Never place cooked foods on the same platter, board or tray that held fresh meats or poultry, unless it has been fully washed and sanitized.

Cook all meat and poultry products to the suggested internal temperature to eliminate any harmful bacteria that may be in the product. I use a meat thermometer to ensure that meats and poultry are cooked to their proper temperature. Your families health and well being are important, so be safe.

I recommend storing leftovers in a shallow covered container and refrigerate. Putting leftovers in the fridge as soon as possible helps to prevent bacteria from growing.

As a child growing up, my Mother had a great way of dealing with leftovers that may have been spoiled or safe to eat. She would say, **"When in doubt, throw it out."**

"Food is a part of our contract
with life." - Bryant McGill

GMO's Pros & Cons

Genetically modified foods (GMO) have been shown to cause harm to humans, animals, and the environment. Despite societies growing awareness and opposition to GMO's, more and more foods continue to be genetically altered.

What is a GMO? A genetically modified organism (GMO) is any organism whose genetic makeup has been altered using genetic engineering techniques.

Currently there is no regulation anywhere in the United States requiring the disclosure of GMO ingredients to consumers. Both Europe and South America highly regulate the cultivation and sale of genetically modified crops and often require the labeling of foods that contain GMO ingredients. The ongoing debate about the effects of GMOs on health and the environment and whether GM food in the U.S. should be labeled is still a controversy in our Country.

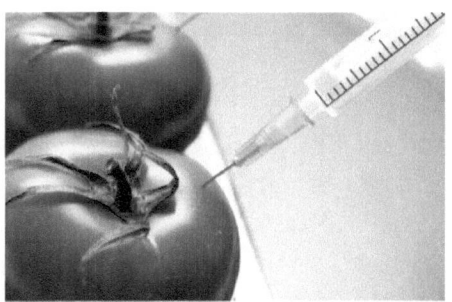
CNN Photo

Some people feel that GMOs are perfectly safe

and there is no evidence to suggest any long-term risk from their consumption. Meanwhile others argue that we are playing God with nature. There has been some scientific evidence suggesting potential health risks caused by GMO's. Scientists with the FDA have warned that GMO foods can create unpredictable, hard-to-detect side effects, including allergies, toxins, new diseases, and nutritional problems.

So, who benefits from GMO foods? The main benefactor of GMO's are farmers, because GMOs, increase farming efficiency. GMOs increase resistance to crop diseases, increase crop yield, growing time, and are often able to survive in harsh weather conditions. GMOs can improve animal health by increasing the animals' hardiness, resistance to disease, and their feed efficiency. GMOs reduce the chemicals needed for crop protection. GMOs have already led to more nutritious staple foods. Finally, a GMO technique nicknamed "pharming" helps by creating plants engineered to produce vaccines, proteins and other pharmaceutical products to help our society.

For every positive in life, there is a negative. According to Brown University, "GMO foods can present significant allergy risks to people." Some GMO foods have had antibiotic features built into them to make them immune or resistant to diseases or viruses, according to Iowa State University. The university also warns that ingestion of GMO foods and regular exposure to antibiotics may be contributing to the decreased effectiveness of antibiotic drugs. A risk is that modified genes may escape into the wild and a super weed that is resistant to herbicides can be created, according to Brown University.

The choice of whether you use or avoid GMO's in your life is a personal decision. I personally try to avoid GMO foods. My biggest concern is that there has not been enough testing of GMO has and no very long term testing to detect possible problems.

My advice for avoiding GMO's is to cook at home using certified organic ingredients. When buying produce such as corn, look for the PLU code on the label. If it starts with the number 9, it is organic. If you use packaged or prepared foods, look for "USDA Certified Organic" on the label. Packaged foods that contain at least 70 percent organic ingredients are not permitted by the USDA to contain GMO's. I personally assume that all restaurants are serving GMO's, unless stated otherwise on their menu.

My Top 10 Worst GMO Foods List

Aspartame - created with genetically modified bacteria.

Canola - obtained from rapeseed through a series of chemical actions.

Corn - resistant to rootworms, drought, and pests.

Cotton - modified to adapt to climate and local factors, resist pests, herbicides, viruses, and fungus.

Dairy - growth hormones

Papaya - modified to resist viruses

Soy - modified to resist pathogens, be tolerant to drought and salinity, resist pests, and fungus.

Sugar - tolerant to drought and to resist herbicides, viruses, and fungus.

Yellow Squash - modified to resist viruses

Zucchini - modified to resist viruses

I recently wrote a book titled, **The Complete Gluten Free Cookbook** and was asked if, "Is genetically modified wheat causing increases in gluten issues?" The answer after much research is "No". For wheat to be considered genetically modified, it needs to have its genome altered through gene splicing in a laboratory.

My research showed that wheat has changed over the last few centuries. This process of change is called hybridization. Many scientists feel these changes could be one of the causes of societies increased inability to tolerate gluten. In hybridization, scientists do not change a plant's genome. Instead, they choose strains of a plant with desirable qualities, and breed them to reinforce those qualities. Over time, successive generations of a particular plant can look very different from the plant's ancestors.

Dr. William Davis, author of the anti-wheat best-selling book **Wheat Belly** states, "Small changes in wheat protein structure can spell the difference between a devastating immune response to wheat protein versus no immune response at all".

The best and safest wheat is einkorn. It is the purest and most ancient form of wheat available as it only has 14 chromosomes and is naturally very low in gluten! Today's modern wheat has 42 chromosomes thanks to mans meddling with Mother Nature.

My answer to whether GMO's are safe for our consumption is simple, **"Steer clear of GMO's altogether."**

Terminology

Air-conditioning - equipment for treating air to control simultaneously its temperature, humidity, cleanliness and distribution to meet the requirements of a conditioned space.

Biodegradable: materials that decompose when naturally occurring organisms, like bacteria and earthworms, feed on them.

CFL's - compact fluorescent light bulbs are more energy efficient than standard incandescent light bulbs and last longer.

Compost - material that breaks down to become what is effectively dirt. It contains no toxins and can support plant life.

Global Warming - increase in the earth's average temperature that causes changes in climate patterns worldwide.

Energy Star - a federal program that labels household products that have met energy-efficient standards set by the U.S. Environmental Protection Agency and the U.S. Department of Energy.

Geothermal - a method of heating and cooling a building using underground heat or electricity generated from naturally occurring geological heat sources.

GMO - any organism whose genetic makeup has been altered using genetic engineering techniques.

Greenhouse Gases - gases like CO_2 that trap heat in the earth's atmosphere.

Hydroelectric - energy gathered from the force of moving water.

Indoor Air Quality (IAQ) - measurement of the cleanliness of indoor air based on the amount of particles, dust, mold, gases, and pollutants present.

Lighting - the act of igniting or illuminating.

Low Flow - plumbing fixtures include faucets, toilets and showerheads.

Mold - growth of minute fungi forming on vegetable or animal matter, commonly as a downy or furry coating, and associated with decay or dampness.

NASA - National Aeronautics and Space Administration (NASA) is an agency of the United States government, responsible for the nation's public space program.

Non-potable Reuse - reused water that is not used for drinking, but is safe to use for irrigation or industrial purposes.

Ozone - is a molecule of three oxygen atoms bound together. It is unstable and highly reactive. Ozone is used as a bleach, a deodorizing agent, and a sterilization agent for air and drinking water. At low concentrations, it is toxic.

Postconsumer Material - products that have been thrown away and that some people are turning into new products.

Potable Water - drinking water.

Potable Reuse - reused water you can drink.

Recycling - to undergo reuse or renewal be subject to or suitable for further use, activity.

Renewable Resources - resources that are always being replenished through natural processes, even as they are being consumed like trees, water, and sunlight.

Reused Water - water used more than once or recycled.

Solar Energy - energy derived from the Sun's radiation.

Volatile Organic Compounds (VOCs) - chemical compounds that have high enough vapor pressures under normal conditions to vaporize and enter the atmosphere—a process called "off gassing."

Water - transparent, odorless, tasteless liquid, a compound of hydrogen and oxygen.

Wind Energy - defined as the "power generated by harnessing the wind, usually by windmills.

Wind Farm - a large group of wind-driven generators for electricity supply.

Resources

Approach Odor Eliminator
1117 East Putnam Avenue
Riverside, CT 06878
www.approachit.net

ASHRAE
1791 Tullie Circle, N.E.
Atlanta, GA 30329
www.ashrae.org

Aurora Home Services
110 Chestnut Ridge Rd Suite 200
Montvale, NJ 07645
www.aurorahomeinspections.com

Britannica Encyclopedia
331 North La Salle Street
Chicago, IL 60610
www.Britannica.com

Brown University
Providence, RI 02912
www.brown.edu

Burt's Bees, Inc.
P.O. Box 24305
Oakland, CA 94623-1305
www.burtsbees.com

Centers for Disease Control (CDC)
1600 Clifton Road
Atlanta, GA 30333
www.cdc.gov

Citra-Solv
P.O. Box 2597
Danbury, CT 06813
www.citra-solv.com

Ecology Center
117 N. Division St.
Ann Arbor, MI 48104
www.ecocenter.org

Environmental Working Group
1436 U St. N.W. Suite 100
Washington, DC 20009
www.ewg.org

Florida Department of Health
4052 Bald Cypress Way
Tallahassee, FL 32399 -1708
www.floridahealth.gov

Flower Council of Holland
6-8 Catherine St, Salisbury
United Kingdom
Info.uk@flowercouncil.org

Ford Motor Company
P.O. Box 6248
Dearborn, MI 48126
www.ford.com

Green Virgin Products
402 Barbara Lane
Tampa Florida 33609
www.greenvirginproducts.com

Institute of Medicine
500 Fifth Street NW
Washington, DC 20001
www.iom.edu

Iowa State University
Ames, Iowa 50011
www.iastate.edu

Kashi Consumer Affairs
P.O. Box 8557
La Jolla, CA 92038
www.kashi.com

Mold Reporter
www.moldreporter.org

National Cancer Institute
6116 Executive Boulevard
Bethesda, MD 20892
www.cancer.gov

Natural Resources Defense Council
40 West 20th Street
New York, NY 10011
www.nrdc.org

National Science Foundation
4201 Wilson Blvd
Arlington, VA 22230
www.nsf.gov

Natural Society
www.naturalsociety.com

New Jersey Wind
201 King of Prussia Road
Radnor, PA 19087
www.njwind.com

New York Times
www.nytimes.com

SeaYu Enterprises, Inc.
PO Box 16236
San Francisco, CA 94116
www.sea-yu.com

Seventh Generation
60 Lake Street
Burlington, VT 05401
www.seventhgeneration.com

Shaklee
4747 Willow Road
Pleasanton, CA 94588
www.shaklee.com

Sick Building Syndrome
ISBN: 978-1-4303-0531-6
Publisher: Lulu.com
Author: James Hewitt, RN
www.stores.lulu.com/jamesphewitt

The Hartman Group, Inc.
3150 Richards Rd, Suite 200
Bellevue, WA 98005
www.hartman-group.com

The National Alliance on Mental Illness
3803 N. Fairfax Drive, Suite 100
Arlington, Va. 22203
www.nami.org

Toxicological Sciences
Oxford University Press
2001 Evans Road
Cary, NC 27513
www.toxsci.oxfordjournals.org

Toyota Motor Sales
19001 South Western Ave.
Torrance, CA 90501
www.toyota.com

U.S. Department of Labor
Occupational Safety & Health Administration
200 Constitution Avenue
Washington, D.C. 20210
www.osha.gov

US Environmental Protection Agency
1200 Pennsylvania Avenue, N.W.
Washington, DC 20460
www.epa.gov

US National Library of Medicine
8600 Rockville Pike,
Bethesda, MD 20894
www.nlm.nih.gov

Vermont Soap
616 Exchange Street
Middlebury, VT 05753
www.vermontsoap.com

Wikipedia Encyclopedia
200 2nd Ave. South #358
St. Petersburg, FL 33701-4313
www.wikipedia.org

Wild Hibiscus Flower Company
PO Box 246
Richford, VT 05476
www.wildhibiscus.com

World Health Organization
www.who.int/en

Conclusion

First, I would like to thank you for purchasing my book. Writing and creating this book has been part of an educational experience for me. Throughout this book, you have learned all about Living an Eco-friendly Life. I have discussed air quality, energy efficient lights and power, recycling, organic foods, and much more.

Fortunately, every year new information is being discovered and being used by people and society with regard to living and eco-friendly life. This syndrome. This book hopefully has whet your appetite regarding this topic. Sorry, if you expected a book to be filled with scientific data. As I previously mentioned in the Introduction, this book was written in common everyday language for the regular non-medical person.

If you are looking for more information, please check out any of the references I have utilized in this books creation. As time goes by, they will have the newest information available to the public. Increasing a person's knowledge is never a bad thing. Education and knowledge are the best tools you can obtain for use in life.

"An ounce of prevention is worth a pound of cure." - Ben Franklin

About the Author

The Author has led an interesting life. He has worked in a variety of fields ranging from Restaurant Management, Licensed Seaman, Educator, Journalist, and, is currently a Registered Nurse. Academically he has earned a variety of degrees, which are listed in chronological order: AA in Mass Media, BA in Communication Arts, AS in Nursing, MS in Health Care Administration, and finally a PhD in Public Administration. He lives with his wife, Virginia, their two sons, John and Seamus, and 4 dogs in Orange County in New York State.